Ecology and Environment

NIPA® GENX ELECTRONIC RESOURCES & SOLUTIONS P. LTD.
New Delhi-110 034

About the Author

Prof. (Dr.) Abhimanyu Kumar Jha is a Molecular Biologist and has extensive Research experience in Molecular Biology, Animal Biotechnology and Cancer Biology. He is currently working as Professor (Department of Biotechnology) and Dean (School of Biosciences and Technology), Galgotias University, Greater Noida. Before that, he worked as Head of the Department of Biotechnology in Sharda University, Greater Noida. Previously he worked as Dean, Faculty of Life Sciences and Head in Department of Biotechnology, Institute of Applied Medicines and Research, Ghaziabad (U.P). He also worked as Head in Department of Biotechnology, IMS Engineering College, Ghaziabad (U.P), India. He did his Ph.D. from Department of Biotechnology, Punjab University, Chandigarh, India. He has published more than 140 papers in national and international journals of repute and also contributed 48 chapters in books published by Elsevier, Springer and Apple Academic Press. He has published more than 14 books for various subjects and competitive exams which are famous in their domains. He has 14 published patents to his credits. He was awarded an honorary degree of Doctor of Science (D.Sc Honorary) with specialization in Biological Science by The American University, USA. He also received ESDA Senior Scientist Award at 5th World Environment Summit and the Wegrow India Research Excellence Award at International Innovation Research Conclave. His current area of research is on the epidemiology and epigenetics of Oral Squamous Cell Carcinoma, Cervical cancer, Liver Cancer and Neurodegenerative disorders.

He has delivered many keynote lectures, invited talks and chaired the scientific session in several international as well as national conferences. His Research group is also working on Alzheimer's disease and Rheumatoid Arthritis as well as the correlation of contaminants present in groundwater with epigenetic changes in cancer. He also worked on the Natural Compounds-Mediated Reversal of promoter methylation of tumor suppressor genes leading to their reactivation in Cervical Cancer and Acute Lymphoid Leukemia Cell Lines.

Dr. Shatabdi Paul is currently serving as an Assistant Professor in the Department of Biotechnology at Galgotias University, India. She earned her Ph.D. in Biotechnology from the Center of Innovative and Applied Bioprocessing (CIAB), Regional Centre for Biotechnology, Government of India, where she focused on lignin-derived nanomaterials for antimicrobial and anticancer photodynamic therapy. With over 15 peer-reviewed publications in reputed international journals and several book chapters published by Springer and ACS, she has made significant contributions to the field of sustainable functional nanomaterials. Dr. Paul also has postdoctoral research experience as a CSIR Research Associate at NIPER, Mohali, and has previously worked at Banaras Hindu University on breast cancer diagnostics using spectroscopic methods. Her research interests include nanoformulations, biopolymer-based materials, and light-responsive therapeutic platforms for biomedical applications.

Runjhun Mathur is currently working towards her PhD at Dr. A.P.J. Abdul Kalam Technical University, Lucknow. She has published more than 30 papers in national and international journals of repute. She has also contributed 8 chapters in books published by Springer and few chapters in Biology Foundation for Class X, published by Universities Press. She has also contributed few chapters to the book " Cell Biology- The Fundamentals". She is one of the Co-author for the book "Biology for NEET" (both Volume I and II) published by Universities Press. She worked as a lecturer at Mahatma Jyoti Rao Phoole Women's College in Jaipur for three years. She obtained her MSc in Biotechnology from the University of Rajasthan, Jaipur.

Ms. Aishwarya holds a Master's degree in Biotechnology from Gautam Buddha University, Greater Noida, She is presently employed as a scientific writer. She has been recognized with an Excellence Award in English Speaking Skills by the American Institute of English Speaking. Her linguistic capabilities are extensive, as she is fluent in French, German, English, and Hindi. She has furthered her proficiency in French by obtaining a Certificate of Proficiency from Chaudhary Charan Singh University, Meerut. With over two years of experience in teaching, Ms. Aishwarya is also committed to social service. During her undergraduate years, she actively participated in the National Service Scheme (NSS), an initiative supported by the Government of India, Ministry of Youth Affairs and Sports. Presently, she is also working as a Rakshin, a role focused on preventing child sexual abuse. Furthermore, she has earned a six-year Diploma in Kathak from Prayag Sangeet Samiti, Allahabad.

Ecology and Environment Principles, Issues and Sustainability

Abhimanyu Kumar Jha
Shatabdi Paul
Runjhun Mathur
Aishwarya

NIPA® GENX ELECTRONIC RESOURCES & SOLUTIONS P. LTD.
New Delhi-110 034

NIPA® GENX ELECTRONIC
RESOURCES & SOLUTIONS P. LTD.

101,103, Vikas Surya Plaza, CU Block
L.S.C.Market, Pitam Pura, New Delhi-110 034
Ph : +91 11 4386 0225, 9717133558, 9540816132
E-mail: newindiapublishingagency@gmail.com
Website: www.niparesources.com

Print ISBN: 978-93-58879-24-7

ebook ISBN: 978-93-5887-631-4

Composed and Designed by NIPA®.

To

Lord Shiva, Mata Parvati and their family

and

My parents, Sh. Arjun Kumar Jha and Smt. Subhadra Jha

—Abhimanyu Kumar Jha

Acknowledgements

We would like to express our gratitude and acknowledge the people who have contributed most to the success of this book. We would like to express our heartfelt gratitude to the management of Galgotias University, Greater Noida, especially Shri. Suneel Galgotia, esteemed Chancellor; Hon'ble Dr. Dhruv Galgotia, CEO; and the dynamic Ms. Aradhana Galgotia, Director of Operations, for providing constant motivation and support. We also extend our sincere thanks to Dr. Ankush Mittal, Vice Chancellor, for his invaluable guidance and encouragement.

Further, we would like to take this opportunity to express our gratitude to our parents and family for their endurance, love, wishes and support, which helped us in completing the work. In particular, Dr. Abhimanyu Kumar Jha would like to thank his parents, Sh. Arjun Kumar Jha and Smt. Subhadra Jha, his wife Mrs. Meenakshi Jha and their son, Master Aayushman Jha. Dr. Abhimanyu Kumar Jha would also like to acknowledge the blessings and support of his in-laws. We also want to thank and appreciate all our family members who offered their help willingly.

We express our gratitude to the student audience over the past few years, and our publisher, NIPA Genx Electronic Resources & Solutions Pvt Ltd., for giving us the opportunity to serve this audience. We want to convey our admiration and gratitude to the scientific community for everything they have done and are now doing to provide us with the best scientific advancements.

Above all, we want to express our gratitude to the Supreme Power, who has blessed us abundantly.

Abhimanyu Kumar Jha
Shatabdi Paul
Runjhun Mathur
Aishwarya

Preface

The growing environmental challenges of the 21st century—ranging from climate change, loss of biodiversity, pollution, to unsustainable resource use—underscore the urgency of ecological awareness and action. This book, an integrated compendium on Ecology and Environmental, has been crafted with the objective of providing students, educators, and concerned citizens a comprehensive understanding of the natural world and the complex interactions that sustain life on Earth.

The content presented herein is drawn from a multidisciplinary approach, encompassing core concepts in biology, chemistry, geography, environmental science, and public policy. Beginning with fundamental definitions and concepts, the chapters progressively explore intricate themes such as environmental degradation, pollution, waste management, energy flow, biogeochemical cycles, population dynamics, and species interactions. Special attention has been given to environmental issues pertinent to India, providing a regional context to global problems.

This book aims to empower its readers with the knowledge needed not just to understand environmental systems, but to evaluate human impact critically and propose sustainable solutions. In doing so, it bridges the gap between theoretical insights and practical applications—emphasizing the importance of responsible action, conservation strategies, and scientific inquiry.

Whether you are a student exploring ecology for the first time or a teacher seeking classroom resources, this volume is structured to support inquiry-based learning through summaries, review questions, multiple-choice items, and thought-provoking discussions. It is our hope that this work serves as a catalyst for fostering environmental literacy, advocacy, and a deeper sense of stewardship toward the planet.

We are deeply grateful to the scholars, researchers, and educators whose contributions and insights have shaped this compilation. May this book inspire curiosity, critical thinking, and, most importantly, commitment to a greener, healthier future.

Authors

Contents

Section VI: Environmental Governance and Global Perspectives

Section I
Foundations of Ecology and Environment

1

Understanding the Environment

The term "Environment" originates from the French word "Environner," signifying the encircling or surrounding elements. It encompasses a multitude of factors that encompass both humanity and other living organisms. Environmental studies delve into the intricate interplay between organisms, their surroundings, and the myriad factors influencing life on Earth, such as atmospheric conditions, food chains, and the water cycle. Given its fundamental role in understanding our planet's dynamics, this field holds significance for individuals across all walks of life.

The scope of environmental studies is extensive and multifaceted, catering to the needs of not just scholars but also the general populace.

1. Promoting awareness about the region's renewable and non-renewable resources, analysing their distribution, utilization patterns, and ensuring their sustainable management.
2. Providing insights into ecological systems and their cause-and-effect relationships.
3. Offering essential information regarding biodiversity and the potential threats to various species in the environment.
4. Understanding the causes and impacts of natural disasters and human-induced pollution, along with measures to mitigate their effects.
5. Evaluating different responses to environmental challenges to inform decision-making.
6. Empowering individuals with environmental literacy to make informed decisions and advocate for Earth's protection and enhancement.
7. Shedding light on societal issues like overpopulation, public health, and hygiene, and exploring the role of arts, science, and technology in addressing them.
8. Identifying and fostering eco-friendly practices and technologies tailored to local environmental concerns.

9. Instilling the importance of sustainable resource management for future generations.
10. Bridging the gap between theoretical knowledge and practical applications, emphasizing the diverse uses of the environment while ensuring its preservation.

Importance of Environmental Studies

The significance of environmental studies lies in its holistic approach to understanding various environmental systems, empowering individuals to engage in scientific inquiry and devise practical solutions to contemporary environmental challenges. By fostering the capacity to analyze environmental parameters across aquatic, terrestrial, and atmospheric realms and their interactions with the biosphere and human activities, environmental studies play a crucial role.

Importance

1. Rapid population growth, particularly in developing nations, poses a pressing concern.
2. Earth's finite natural resources are being depleted.
3. Advancements in resource exploitation methods exacerbate overexploitation.
4. Lack of foresight in resource management jeopardizes future generations' access to resources.
5. Uncontrolled exploitation leads to pervasive pollution.
6. Environmental degradation adversely impacts global health.
7. Collective responsibility is imperative to address environmental degradation.
8. Education and training are essential to preserve biodiversity and prevent species extinction.
9. Urbanization and industrialization are significant sources of pollution.
10. Expanding protected areas is crucial for wildlife conservation.
11. Environmental understanding necessitates adaptive and sustainable activities.
12. Encouraging community engagement and participation in environmental initiatives is vital.
13. Educational systems should prioritize environmental education.
14. Environmental studies adopt a multidisciplinary approach, integrating various academic disciplines to address environmental issues.

15. Environmental education fosters changes in knowledge, values, behaviors, and lifestyles essential for sustainability and global stability.

Environmental studies encompass a broad spectrum of disciplines aimed at understanding and addressing the myriad factors impacting organisms and their surroundings as shown in table 1. It adopts a multidisciplinary approach to comprehend both the natural world and human influences on its balance. This field serves as an applied science, seeking practical solutions to ensure the sustainability of human civilization within the constraints of Earth's finite resources. Its components span across various fields including biology, geology, chemistry, physics, engineering, sociology, health, anthropology, economics, statistics, and philosophy.

Table 1: Importance of Environmental Studies

Importance Factor	Explanation
Population Growth	Rapid population growth, especially in developing nations, poses a concern.
Resource Depletion	Earth's finite natural resources are being depleted.
Resource Exploitation	Advances in exploitation methods worsen overexploitation.
Foresight in Resource Management	Lack of foresight jeopardizes future access to resources.
Pollution	Uncontrolled exploitation leads to pervasive pollution.
Global Health	Environmental degradation adversely impacts global health.
Collective Responsibility	Collective responsibility is crucial to address degradation.
Education and Training	Essential to preserve biodiversity and prevent extinction.
Urbanization and Industrialization	Significant sources of pollution.
Protected Areas	Expanding protected areas for wildlife conservation.
Adaptive and Sustainable Activities	Environmental understanding necessitates such activities.
Community Engagement	Vital for participation in environmental initiatives.
Environmental Education	Educational systems should prioritize it.
Multidisciplinary Approach	Integrates various disciplines to address issues.
Behavioural Changes	Fosters changes in knowledge, values, behaviours, and lifestyles essential for sustainability.

Major Environmental Issues

Over time, human ambition for unlimited comfort and enjoyment has led to the reckless exploitation of nature's resources, disrupting its inherent stability. This indiscriminate exploitation has given rise to numerous environmental challenges, as mankind's relentless pursuit of resources and dominance over nature disrupts the delicate equilibrium of the environment. Key environmental issues confronting humanity include global warming, water pollution, pesticide pollution, hazardous waste disposal, biomedical waste, electronic waste, and biodiversity loss as shown in Figure 1.

India, now among the top ten industrialized nations globally, boasts a robust industrial infrastructure in core sectors like metals, chemicals, fertilizers, and petroleum. However, this industrial progress has come at a cost, manifesting in the proliferation of pollutants such as pesticides, detergents, plastics, solvents, paints, dyes, and radioactive materials due to advancements in atomic energy. Industrial activities also contribute significantly to environmental degradation through emissions of poisonous gases and the generation of solid waste, exacerbated by mining operations.

The adverse effects of such activities extend to all forms of life within the biosphere, the planet Earth along with its atmosphere sustaining life. Despite the lack of a robust culture of pollution control, there exists a substantial backlog of gaseous, liquid, and solid pollutants in India, including hazardous waste, medical waste, and pesticide residues, alongside issues like automobile pollution, climate change, water pollution, and biodiversity loss.

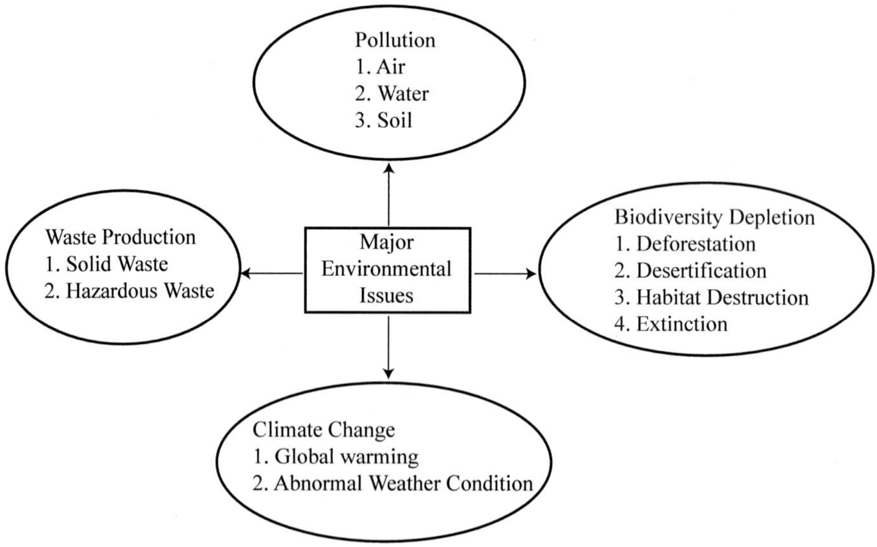

Figure 1: Major Environmental Issues

Industrial Pollution

Among the primary contributors to environmental degradation in urban centers are vehicular and industrial pollution. Since 1975, India's economic growth has been accompanied by a disproportionate increase in industrial and vehicular pollution loads, particularly evident in metropolitan areas like Delhi, where vehicular emissions account for a significant portion of air pollution. The pollutants emitted by vehicles pose various health risks, including respiratory ailments and carcinogenic effects, depending on the type of fuel used, primarily diesel or petrol.

Climate Change

The escalation in concentrations of anthropogenic greenhouse gases (GHGs) in the atmosphere, including carbon dioxide (CO_2), methane (CH_4), and nitrous oxide (N_2O), has been notable since the late 19th century. As outlined in the Third Assessment Report (TAR) of the Intergovernmental Panel on Climate Change, this surge in GHG concentrations, such as a 29 percent increase in CO_2, a 150 percent rise in CH_4, and a 15 percent uptick in N_2O over the past century, has corresponded with a global mean surface temperature rise of 0.4–0.8°C. Concomitantly, precipitation patterns have become more sporadic, while the frequency and intensity of extreme weather events have heightened. Additionally, sea levels have seen an average annual rise of 1–2 mm during this period. Continual GHG concentration increases are poised to precipitate climate change, potentially causing significant ecosystem alterations and posing grave disruptions to livelihoods, economies, living conditions, and public health. The United Nations Framework Convention on Climate Change mandates that parties safeguard the climate system based on their "common but differentiated responsibilities" and respective capacities. In 1990, developed nations (including Australia, Canada, the USA, Europe, the former USSR, and Japan) accounted for roughly 66 percent of total global GHG emissions, a figure that decreased to 54 percent by 2000, albeit countered by escalating emissions from China. Conversely, the South Asian region, with India contributing the lion's share, only accounted for 3 percent of global GHG emissions in 1990, with this proportion experiencing a marginal 4 percent increase by 2000.

Water Pollution

Regarding water pollution, India boasts 12 major rivers spanning a total catchment area of 252.8 million hectares. Domestic sources contribute approximately 75 percent of wastewater, with most urban areas lacking adequate sewage treatment infrastructure, and rural locales often devoid of such facilities altogether. As per data from the Central Pollution Control Board, out of 8,432 large and medium-scale industries nationwide, only 4,989 had implemented proper wastewater treatment measures by 2000. Among over two million small-scale industrial units, many, including highly polluting sectors like tanneries, lack any treatment facilities, leading to untreated waste discharge into the country's water bodies.

Pesticides Poisoning

Pesticide poisoning poses a grave threat to farmers and workers, impacting an alarming number of individuals worldwide. Each day, around 68,000 farmers and workers suffer from pesticide poisoning, with an estimated 25 million affected annually. Particularly in developing regions like Asia, farmers and agricultural workers face direct exposure to pesticides during mixing and spraying activities. Globally, approximately 3 million people are poisoned by pesticides annually, resulting in 200,000 deaths.

Beyond the acute cases reported, the chronic long-term effects of pesticide exposure are increasingly concerning. These effects include various cancers, adverse impacts on specific body organs and systems, as well as disruptions to the endocrine system, leading to issues such as reduced sperm count, reproductive organ abnormalities, and increased incidences of breast cancer. Communities and consumers face a hidden risk of pesticide exposure through soil, air, and water contamination. The chronic effects are particularly alarming as recent studies have linked certain pesticides to cancer, reduced fertility, disruption of the endocrine system, and suppression of the immune system.

Global Stories of Pesticide Poisoning and Resistance

- The struggles of plantation workers in Malaysia against pesticides like Parquet, highlighting their fight for workers' rights.
- Former workers of the International Rice Research Institute (IRRI) in the Philippines poisoned by pesticides used in IRRI test fields, with inadequate support from the institute. Additionally, the fisher community in Kamukhaan, Philippines, faced pesticide poisoning and environmental devastation due to neighbouring banana plantations' pesticide use.
- Communities in Kasar code, Kerala, poisoned by aerial spraying of Endosulfan by a plantation corporation in Kerala, India.
- Farming communities in Warangal, Andhra Pradesh, affected by pesticide poisoning during spraying. Warangal is already notorious for high rates of cotton farmer suicides, partly attributed to pest resistance to pesticides.

The Pesticides Action Network Asia and the Pacific (PANAP) initiated "No Pesticide Use Day" in 1998 to protest global pesticide manufacturing and usage. This day commemorates the victims of the 1984 Bhopal Disaster, serving as a stark reminder of the ongoing suffering caused by chemical pesticide contamination.

Pesticides in Soft Drinks

Soft drinks, which are non-alcoholic, water-based beverages often flavored, sweetened, and carbonated, have become ubiquitous in the market, with major players like PepsiCo and Coca-Cola dominating the industry in India. However, a laboratory report conducted by CSE in 2003 revealed alarming findings regarding pesticide contamination in soft drinks sold in India. The report uncovered high levels of toxic pesticides and insecticides, posing serious health risks including cancer, damage to the nervous and reproductive systems, birth defects, and immune system disruption. Notably, both Coca-Cola and Pepsi had similar concentrations of pesticide residues. In contrast, soft drink brands tested in the US showed no pesticide contamination, suggesting a double standard in manufacturing practices.

In terms of regional variations, Kolkata emerged with the highest pesticide content in soft drinks among 18 cities surveyed, with levels reaching about 51.7 parts per billion (ppb), while Guwahati had the lowest levels. These pesticides, when consumed, can inflict irreversible harm on the human body.

Hazardous Waste

Hazardous waste, whether in liquid, solid, or gas form, poses significant risks to human health and the environment when not managed properly as shown in Figure 2. In India, the generation of hazardous waste amounts to 6-7 million tonnes annually, with variations depending on the type and quantity of waste produced. Major contributors to hazardous waste in India include industries dealing with petrochemicals, pharmaceuticals, pesticides, paints, dyes, fertilizers, and chlor-alkali. The lack of a proactive approach to waste management has led to the escalating generation of hazardous waste, presenting a serious challenge in India. Implementation of regulations regarding hazardous waste management has been negligent, resulting in a regular influx of such waste into the country. Additionally, India often imports hazardous waste under the guise of reuse and recycling, despite a lack of environmentally friendly technologies for such purposes.

Figure 2: Hazardous Waste

Indiscriminate generation, improper handling, storage, and disposal of hazardous waste are key factors contributing to adverse environmental and human health impacts. There is an urgent need to shift from traditional pollution control measures to strategies focused on pollution prevention, waste minimization, cleaner production, and toxics reduction.

Biomedical Waste

Biomedical waste, encompassing both organic and inorganic materials generated by healthcare facilities, presents another significant challenge. Hospitals generate substantial amounts of waste daily, with a portion being infectious or hazardous. The improper disposal of medical waste poses serious risks, including the potential for cross-infection among patients. While regulations exist for managing medical waste, effective training of healthcare staff and the establishment of comprehensive waste management systems in hospitals are urgently needed. Plastics constitute a significant portion of medical waste due to their widespread use in healthcare settings. While plastics reduce the risk of infection transmission within hospitals, challenges related to their use and disposal persist. Mercury, a highly toxic substance, poses greater dangers than lead and arsenic. Efforts to reduce the proliferation of hazardous waste and enhance waste management practices are crucial to mitigating environmental and health risks associated with biomedical waste and other hazardous materials.

Cracking down of Crackers

Diwali, once celebrated with joy and festivity, has unfortunately transformed into a festival marked by pollution, particularly due to noise, crackers, artificially colored sweets, and associated health hazards. During this time, cities become inundated with toxic fumes and gases such as carbon dioxide (CO_2), sulfur dioxide (SO_2), nitrogen dioxide (NO_2), and suspended particulate matter (SPM), turning the air into a hazardous mix. Among the worst affected by this pollution are children, pregnant women, and individuals with respiratory issues. Additionally, the factories producing crackers often flout safety regulations and exploit child labor, with children enduring long hours of work in unsafe conditions for minimal wages. Exposure to chemicals in these environments poses severe health risks, leading to diseases affecting the lungs, kidneys, skin, and eyes.

E-Waste

E-waste, comprising discarded electronic products nearing the end of their useful life, presents another significant environmental challenge. Computers are typically replaced every two to four years, while cell phones have a life-cycle of less than two years in industrialized nations. These electronic devices contain various hazardous materials, with computer screens alone containing approximately 20% lead by weight. Informal disposal of e-waste, including large household appliances like refrigerators, air conditioners, computers, stereo systems, and mobile phones, contributes to an annual volume increase of 3-5%. Major pollutants in e-waste include heavy metals such as mercury (Hg), lead (Pb), cadmium (Cd), chromium (Cr VI), as well as flame retardants like polybrominated biphenyls (PBB) and polybrominated diphenyl ethers (PBDEs).

Globally, e-waste generation ranges from 20 to 50 million tonnes annually, with the USA accounting for 1% to 3% of total municipal waste generation. In the EU, e-waste generation stands at 5 to 7 million tonnes per year, or about 14 to 15 kg per capita, and is projected to grow at a rate of 3% to 5% annually. In developed nations, e-waste currently comprises 1% of total solid waste generation, expected to increase to 2% by 2010. Efforts to address e-waste management and reduce its environmental impact are crucial to mitigating the growing problem of electronic waste pollution.

Magnitude of the Issue in India

1. E-waste generation in India has surged from 146,000 tonnes to 470,000 tonnes by 2011, with projections indicating a further increase to over

800,000 tonnes by 2012. This represents a growth rate of approximately 15%.

2. Sixty-five cities in India contribute to more than 60% of the total e-waste generation, with major metropolitan areas like Mumbai, Delhi, Bangalore, Chennai, Kolkata, Ahmedabad, Hyderabad, Pune, Surat, and Nagpur being the top contributors.
3. Illegal imports of e-waste amount to around 50,000 MT per year in India.

Table 2: Environmental Pollutants in India

Pollutant Category	**Key Pollutants**	**Sources**
Air Pollution	CO_2, SO_2, NO_2, SPM	Vehicles, industrial emissions, burning of fossil fuels.
Water Pollution	Industrial waste, domestic sewage	Factories, urban runoff, lack of sewage treatment infrastructure.
Pesticide Pollution	Pesticides (various types)	Agricultural activities, improper pesticide use and disposal.
Hazardous Waste	Chemicals, heavy metals	Petrochemical industries, pharmaceuticals, pesticides, paints, dyes.
Biomedical Waste	Infectious waste, plastics, mercury	Healthcare facilities, hospitals.
E-Waste	Heavy metals (Hg, Pb, Cd, Cr VI), flame retardants	Discarded electronic products, informal disposal methods.

Loss of Biodiversity

The ongoing loss of biodiversity due to factors like overexploitation, habitat degradation, deforestation, and land pollution poses a significant threat to global ecosystems and human existence. If current trends persist, it is estimated that a quarter of the world's species may become extinct by 2050. The rate of species extinction, which historically averaged one per year over the past 600 million years, has accelerated to potentially dozens per day in the present era. Thus, the conservation of biodiversity has become a critical environmental imperative, requiring concerted efforts from nations, government agencies, organizations, and individuals to protect and enhance biological diversity while meeting the needs for natural resources.

Remaining Challenges

There is a need for improvement in monitoring and verification protocols for carbon sequestration in soil-plant ecosystems to facilitate quantitative economic and policy analyses. These protocols must be universally acceptable to scientists, policymakers, landowners, and business groups, as well as practical for use by government agencies and licensed professionals. Additionally, techniques are required to quantify the overall net beneficial impact of agricultural and silvicultural practices on all greenhouse gases, including methane (CH_4) and nitrous oxide (N_2O), while recognizing and evaluating other beneficial services derived from improved land practices, such as changes in soil quality, productivity, water and air quality, and erosion. Furthermore, recommended carbon sequestration practices should demonstrate long-term stability, necessitating long-term studies to ensure the effectiveness of current strategies over extended periods (at least 20-50 years) from a whole ecosystem accounting perspective.

Summary /Key points

1. Industrial Pollution: Significant in urban centers, exacerbated by vehicular emissions and industrial activities.
2. Climate Change: Rising greenhouse gas concentrations leading to global warming, altered precipitation patterns, extreme weather events, and sea-level rise.
3. Water Pollution: Major rivers polluted due to inadequate sewage treatment and industrial waste discharge.
4. Pesticide Poisoning: Acute and chronic health effects on farmers and communities, highlighting the need for better pesticide management.
5. Hazardous Waste: Includes petrochemicals, pharmaceuticals, pesticides, and lack of proper waste management.
6. Biomedical Waste: Generated by healthcare facilities, posing cross-infection risks and needing better management systems.
7. E-Waste: Growing concern with discarded electronics containing hazardous materials, necessitating improved disposal and recycling practices.
8. Biodiversity Loss: Resulting from habitat degradation and overexploitation, threatening global ecosystems.

Very Short Answer Type Questions

1. What is the origin of the term “Environment”?
2. What is the primary focus of environmental studies?
3. Name one significant environmental challenge posed by industrial pollution.

4. What are some major pollutants found in soft drinks according to the CSE report?
5. How does hazardous waste pose a threat to human health and the environment?
6. What is the significance of e-waste management in India?
7. What are the key factors contributing to biodiversity loss?
8. How does overpopulation impact environmental sustainability?
9. What are the main pollutants emitted by vehicles contributing to air pollution?
10. What is the primary objective of the "No Pesticide Use Day" initiative?

Long Answer Type Questions

1. How has Diwali transformed into a festival of pollution?
2. What are the major findings of the CSE report on pesticide contamination in soft drinks?
3. How does industrial pollution affect urban areas in India?
4. What are the implications of climate change for global ecosystems and human livelihoods?
5. Describe the challenges associated with wastewater treatment in India.
6. What are the health risks associated with pesticide poisoning?
7. How does the improper disposal of medical waste impact public health?
8. Why is e-waste management important for environmental sustainability?
9. How has industrialization contributed to environmental degradation in India?
10. Discuss the significance of biodiversity conservation in mitigating environmental crises.

Long Answer Type Questions

1. Explain the significance of environmental studies in addressing contemporary environmental challenges.
2. Discuss the findings of the CSE report on pesticide contamination in soft drinks and its implications for public health.
3. Analyse the impact of industrial pollution on urban areas in India, considering both environmental and public health aspects.
4. Evaluate the effectiveness of current waste management practices in addressing hazardous waste disposal in India.

5. Explore the relationship between climate change and global environmental sustainability, highlighting key mitigation strategies.
6. Assess the role of government policies and regulations in promoting sustainable waste management practices in India.
7. Discuss the challenges associated with water pollution in India and propose measures to address them.
8. Examine the socioeconomic implications of pesticide poisoning on farming communities in India.
9. Investigate the potential health hazards associated with biomedical waste mismanagement and propose solutions.
10. Analyse the impact of e-waste generation on environmental sustainability and propose strategies for effective e-waste management in India.

Multiple Choice Questions

1. What does the term "environment" originate from?

   A. Latin word "Environ" **B. French word "Environner"**

   C. Greek word "Environos" D. German word "Environ"

2. Which of the following is NOT a primary objective of environmental studies?

   A. Promoting awareness about regional resources

   B. Understanding economic growth models

   C. Providing insights into ecological systems

   D. Offering information about biodiversity

3. Which major environmental issue is associated with the rise in global temperatures due to greenhouse gases?

   A. Water Pollution B. Hazardous Waste

   C. Global Warming D. Biodiversity Loss

4. What percentage of wastewater in India is contributed by domestic sources?

   A. 50% **B. 75%**

   C. 90% D. 25%

5. Which of the following pollutants is NOT typically associated with e-waste?

   A. Mercury (Hg) B. Lead (Pb)

   C. Sulfur Dioxide (SO_2) D. Cadmium (Cd)

6. What annual increase in sea levels has been observed during the 20th century due to climate change?

   A. 1-2 mm B. 3-4 mm

   C. 5-6 mm D. 7-8 mm

7. Which environmental issue is most directly linked to the agricultural use of pesticides?
   A. Global Warming
   B. Hazardous Waste
   C. Pesticide Poisoning
   D. E-Waste
8. Which Indian city was reported to have the highest pesticide content in soft drinks according to a 2003 laboratory report?
   A. Delhi
   B. Mumbai
   C. Kolkata
   D. Chennai
9. What is the significance of "No Pesticide Use Day"?
   A. Celebrating the benefits of pesticides
   B. Commemorating the Bhopal Disaster victims
   C. Promoting increased pesticide usage
   D. Launching new pesticide products
10. Which pollutant is a major concern in biomedical waste due to its toxicity?
    A. Lead (Pb)
    B. Mercury (Hg)
    C. Cadmium (Cd)
    D. Chromium (Cr)
11. Which country accounts for the highest percentage of global greenhouse gas emissions as of 2000?
    A. USA
    B. China
    C. India
    D. Russia
12. Which practice is emphasized to ensure the sustainability of human civilization within Earth's finite resources?
    A. Resource Exploitation
    B. Overpopulation
    C. Sustainable Resource Management
    D. Uncontrolled Urbanization
13. What is the estimated annual global e-waste generation range?
    A. 10-20 million tonnes
    B. 20-50 million tonnes
    C. 50-100 million tonnes
    D. 100-150 million tonnes
14. Which element is highlighted as having higher toxicity compared to lead and arsenic in biomedical waste?
    A. Cadmium
    B. Mercury
    C. Chromium
    D. Sulfur
15. The rapid population growth is a significant concern primarily in which regions?
    A. Developed nations
    B. Industrialized regions
    C. Developing nations
    D. Urban centers

16. Which practice should be prioritized to reduce environmental and health impacts of hazardous waste?

    A. Pollution control measures
    B. Waste minimization and cleaner production
    C. Increased industrial activity
    D. Expanding urban areas

17. The improper disposal of which type of waste poses serious risks including potential for cross-infection among patients?

    A. Hazardous Waste
    B. Biomedical Waste
    C. E-Waste
    D. Pesticide Waste

18. What major environmental challenge arises due to the burning of firecrackers during Diwali?

    A. E-Waste
    B. Air and Noise Pollution
    C. Water Pollution
    D. Biodiversity Loss

19. Which Indian cities contribute the most to the country's e-waste generation?

    A. Delhi and Mumbai
    B. Bangalore and Chennai
    C. Kolkata and Hyderabad
    D. All of the above

20. What approach does environmental studies adopt to understand and address environmental issues?

    A. Singular scientific discipline
    B. Multidisciplinary approach
    C. Economic-only perspective
    D. Technology-focused analysis

2

Basics of Environment

Environmental studies is a multidisciplinary science encompassing branches like chemistry, physics, life science, and more. It focuses on understanding physical phenomena in the environment, including the sources, reactions, transport, and effects of biological species in air, water, and soil. It also examines the impact of human activities on these aspects.The totality of an organism's surroundings, encompassing both living and non-living elements that influence its existence, is termed its environment. This environment includes natural components like land (often referred to as the lithosphere), air (the atmosphere), and water (the hydrosphere), along with the diverse communities of organisms that share this space. The term itself originates from the French word 'environner,' meaning to encircle or surround. Initially, the concept of environment solely encompassed the physical aspects of our planet Earth. This included the landmasses, the atmosphere, and the water bodies, along with the various living organisms that thrived within them. However, with the advancement of society and human civilization, the definition of environment has expanded to encompass the social, economic, and political spheres of human existence.

At the organism level, the environment plays a crucial role in physiological interactions. This field of study focuses on how different organisms adapt to their surroundings, not just for survival but also for successful reproduction and population growth. All living things, from the microscopic viruses to human beings, depend on the environment to fulfill their essential needs, such as sustenance (food), shelter, water, and oxygen. In simpler terms, an environment can be understood as the sum total of interactions between water, air, and land. These interactions occur not only amongst themselves but also with human beings, other living organisms, and the various material goods we produce.

It encompasses both the physical and biological surroundings, along with the intricate connections that exist between them. The field of environmental studies equips us with the tools to understand the intricate workings of our planet's environment. It sheds light on how human activities impact the environment and vice versa, fostering a more balanced relationship between humans and the natural world. The very nature of the environment is inherently universal, encompassing a vast array of disciplines. These include physics, chemistry, geology, geography, history, economics, physiology, biotechnology, remote sensing, geophysics, soil science, and hydrology etc.

Importance of the Environment

The environment encompasses all living (biotic) and non-living (abiotic) components that interact and influence life on Earth. This intricate web of elements is essential for the survival of all organisms. Consequently, environmental issues like global warming, ozone depletion, deforestation, resource depletion, and biodiversity loss have far-reaching consequences, impacting everyone. Studying environmental processes in the hydrosphere (water), atmosphere (air), lithosphere (land), and biosphere (living organisms) helps us understand how human activities can pollute the environment. Environmental science also plays a crucial role in establishing benchmarks for healthy and sustainable ecosystems.

Defining Our Environment

The environment can be understood as the surroundings or setting in which all living things exist and interact. It includes both the natural forces and conditions that shape life, such as climate, geography, and available resources.

Various Definitions of Environment

1. **Douglas and Holland:** Environment refers to the entirety of external forces, influences, and conditions that affect the life, behavior, growth, development, and maturity of living organisms.
2. **C.C. Park:** Environment is the sum total of all conditions surrounding an organism at a specific point in space and time.
3. **Encyclopedia Britannica:** Environment encompasses all external influences acting on an organism, including physical, biological factors, and interactions with other organisms.
4. **U.S. Council on Environmental Quality:** The total environmental system includes not just the biosphere but also how humans interact with their natural and man-made surroundings.
5. **K.R. Dikshit (1984):** A comprehensive definition of the environment views it as a holistic system with various spatial, elemental, and socioeconomic components. These components are distinguished by their quality, characteristics, and how living and non-living elements interact within the system.

The Environment: A Complex Web of Interactions

The environment is a unified system comprised of interacting physical, biological, and cultural elements. These elements are intricately linked, influencing each other in countless ways. Physical elements, like landforms, water bodies, climate, and soil, shape the characteristics of human habitats, offering both opportunities and limitations. Biological elements, including plants, animals, microorganisms, and humans, make up the biosphere. Cultural elements, such as economic, social, and political structures, are human-made features that contribute to our cultural environment (Savindra Singh and A. Dubey, 1983).

Figure 1: Components of Environment

A Broad Definition Encompassing Our Surroundings

The environment can be defined in two ways: (1) the physical and social conditions surrounding an organism or group, and (2) the social and cultural factors that influence individuals and communities. Humans inhabit both the natural world and the world we've built, including technological advancements, social structures, and cultural influences. All of these elements contribute to our environment (W.P. Cunningham and M.A. Cunningham, 2004).

Environmental Knowledge: A Multifaceted Discipline

Understanding the environment requires knowledge from various disciplines. Key aspects of environmental knowledge include:

1. Conservation and wise management of natural resources
2. Maintaining and protecting biodiversity
3. Controlling and minimizing environmental pollution
4. Stabilizing human population growth to achieve environmental sustainability
5. Developing alternative sources of renewable energy
6. Strengthening national security through environmental conservation and responsible management

Environmental Issues and Opportunities

Environmental knowledge also addresses critical issues like access to safe drinking water, sanitation, clean air, and fertile land. It promotes the development of sustainable environmental practices, laws, and administration. Additionally, environmental protection, management, and related businesses are emerging as new avenues for protecting and managing our environment.

The Interconnectedness of Life

The relationship between organisms and their environment is complex and multifaceted. No organism can exist in isolation; all living things interact with each other and with the surrounding biotic (living) and abiotic (non-living) elements. Our environment encompasses everything we interact with and rely on for survival. While perspectives on the environment may differ, it's widely accepted that it's a unified system where physical, biological, and cultural elements are intricately interwoven.

Classifying our Environment

Environments can be categorized based on their fundamental structure:

1. **Physical or Abiotic Environment:** This encompasses non-living elements.
2. **Biotic Environment:** This encompasses living organisms.
3. **Cultural Environment:** This refers to human-made features.

The Physical Environment: A Breakdown

The physical environment, also known as the abiotic environment, is further divided based on its physical characteristics and state:

1. **Solid:** The lithosphere, which includes the Earth's solid components.
2. **Liquid:** The hydrosphere, which encompasses water bodies.
3. **Gas:** The atmosphere, which is the gaseous layer surrounding Earth.

These environments can be further classified into smaller units based on location and scale. Examples include mountain, desert, river, and lake environments. Climate also plays a role in shaping the physical environment, creating distinct habitats for biological communities, such as tropical, temperate, and polar environments.

The Biotic Environment: Flora and Fauna

The biotic environment consists of living organisms, including plants (flora) and animals (fauna). Humans are a significant factor within this environment.

Social and Economic Interactions

Organisms form social groups and societies at various levels, leading to the development of a social environment. In this environment, organisms interact with the physical environment to obtain resources for survival and growth. These interactions create an economic environment. It's worth noting that humans are the most skilled and organized social organisms. Human social structures are complex and involve interactions between physical, social, and economic aspects.

Humans and the Environment: Threefold Role

Humans play a unique role in the biotic environment due to their physical, social, and economic characteristics.

1. **Physical Humans:** As biological organisms, humans require basic elements from the physical environment, such as habitat, air, water, and food. They also release waste products back into the ecosystem.
2. **Social Humans:** As social beings, humans establish institutions, organizations, laws, and policies to ensure their survival, well-being, and social welfare.
3. **Economic Humans:** Humans utilize their skills and technologies to extract and utilize resources from both the physical and biotic environments.

This economic function sets humans apart, as they actively transport matter and energy within the ecosystem.

Classifying our Environment

Environments can be categorized based on their fundamental structure:

1. **Physical or Abiotic Environment:** This encompasses non-living elements.
2. **Biotic Environment:** This encompasses living organisms.
3. **Cultural Environment:** This refers to human-made features.

The Physical Environment: A Breakdown

The physical environment, also known as the abiotic environment, is further divided based on its physical characteristics and state:

1. **Solid:** The lithosphere, which includes the Earth's solid components.
2. **Liquid:** The hydrosphere, which encompasses water bodies.
3. **Gas:** The atmosphere, which is the gaseous layer surrounding Earth.

These environments can be further classified into smaller units based on location and scale. Examples include mountain, desert, river, and lake environments. Climate also plays a role in shaping the physical environment, creating distinct habitats for biological communities, such as tropical, temperate, and polar environments.

The Biotic Environment: Flora and Fauna

The biotic environment consists of living organisms, including plants (flora) and animals (fauna). Humans are a significant factor within this environment.

Social and Economic Interactions

Organisms form social groups and societies at various levels, leading to the development of a social environment. In this environment, organisms interact with the physical environment to obtain resources for survival and growth. These interactions create an economic environment. It's worth noting that humans are the most skilled and organized social organisms. Human social structures are complex and involve interactions between physical, social, and economic aspects.

The Environmental Orchestra: A Composition of Abiotic and Biotic Elements

Our planet's environment functions like a complex orchestra, with various components playing crucial roles. These components can be broadly categorized into two main sections: abiotic (non-living) and biotic (living).

The Abiotic Stage: Setting the Scene for Life

The abiotic environment acts as the foundation for life on Earth. It encompasses the physical and chemical elements that influence the distribution and abundance of living organisms. Here are some key players in this abiotic ensemble:

1. **Atmosphere:** This gaseous layer surrounding Earth, extending roughly 300 kilometers above the surface, is essential for life. It's not just composed of gases, but also contains water vapor, dust particles, and even microscopic organisms.
2. **Lithosphere:** Derived from the Greek words for "rocky" and "sphere," the lithosphere is the Earth's outermost solid layer. This layer itself is composed of distinct sub-layers:
   - i. **Core:** Located at the Earth's center, the core is a superheated sphere with a radius of approximately 3,500 kilometers.
   - ii. **Mantle:** Surrounding the core, the mantle is a thick layer roughly 2,900 kilometers deep.
   - iii. **Crust:** The Earth's outermost solid layer, the crust is much thinner than the mantle, averaging around 30 to 70 kilometers thick. It can be further divided into two main types: the basalt-rich oceanic crust and the granitic continental crust.
3. **Hydrosphere:** Encompassing all water on or near Earth's surface, the hydrosphere includes oceans, lakes, rivers, wetlands, ice caps, and even the water vapor in the atmosphere. Interestingly, water exists in all three states of matter on Earth: solid (ice), liquid (water), and gas (water vapour). While over 71% of the planet's surface is covered in water, only a small fraction (around 2.53%) is freshwater.

The Abiotic Stage: Shaping the Conditions for Life

The environment encompasses both physical and biological components, integrating both non-living (abiotic) and living (biotic) elements. These abiotic factors play a critical role in determining where and how well organisms thrive. While they interact, a single factor can sometimes limit an organism's range, acting as a "limiting factor." Here's a breakdown of some key abiotic components:

1. **Physical Factors:** These core elements include temperature, water availability (rainfall), light (energy source), soil properties, and atmospheric pressure.

   i. **Temperature:** This is a fundamental ecological factor, varying significantly across the globe. It generally decreases as we move away from the equator towards the poles and with increasing altitude. Some organisms, like desert cacti, can tolerate a wide temperature range (eurythermal). Others, like deep-sea creatures, have adapted to thrive within a narrow temperature band (stenothermal).

   ii. **Water:** Essential for life, water availability varies greatly. Deserts showcase adaptations by plants and animals that enable them to survive in arid conditions. In aquatic environments, water quality (chemical composition and pH) becomes crucial. Salinity levels differ in oceans, lakes, and inland waters, with some organisms tolerating a wider range (euryhaline) than others (stenohaline).

Beyond the Basics: Water Quality and Light Variations

This initial categorization provides a foundation, but the abiotic environment is more nuanced. For instance, water quality plays a vital role for aquatic organisms. The chemical composition and pH of water can significantly impact their survival. Additionally, salinity levels vary greatly between oceans, lakes, and inland waters. Some organisms can tolerate a wide range of salinity (euryhaline), while others are restricted to specific salinity ranges (stenohaline).

Light availability is another crucial factor, particularly for autotrophs (organisms that produce their own food) as they rely on sunlight for photosynthesis. Light variations also influence animal behavior. Many creatures use the rhythmic fluctuations in light intensity and duration (photoperiod) as cues for essential activities like foraging, reproduction, and migration. Deep ocean environments, perpetually shrouded in darkness, have fostered unique adaptations in organisms that have learned to thrive without sunlight.

Light and Soil: Illuminating Life's Necessities

Light: Fuelling Life's Engine

Light, a vital source of energy, plays a central role for living organisms, particularly autotrophs. These self-sustaining organisms rely on sunlight to fuel photosynthesis, the specialized process by which they convert light energy into chemical energy for sustenance. Sunlight also influences plant behavior, with many species dependent on the day/night cycle (diurnal variations) and seasonal changes in light intensity and duration (photoperiod) to trigger flowering. Similarly, many animals utilize these light variations as cues for essential activities like foraging for food, reproduction, and migration. While light availability on land is closely linked to temperature (both originating from the sun), the deep ocean environment (beyond 500 meters) presents a stark contrast. Here, perpetual darkness reigns, and organisms have adapted to thrive without the sun's direct influence.

Soil: The Foundation for Terrestrial Life

The properties of soil vary considerably across the globe, shaped by factors like climate (temperature and humidity), weathering processes, and its formation (transported or sedimentary). Physical characteristics such as soil composition, grain size, and structure all influence its ability to hold water, a vital resource for plants. These physical features, along with chemical aspects like pH and mineral content, and the overall topography of an area, largely determine the type of vegetation that can flourish there. This, in turn, dictates the types of animals that can find sustenance and shelter within a particular habitat. Similar principles apply to the sediment characteristics of aquatic environments. The types of bottom-dwelling organisms that thrive in an area are often determined by the specific characteristics of the sediment on the seabed.

Table 1: Structured overview of the various components and interactions within the environment

Component	Description
Environment	The totality of an organism's surroundings, encompassing both living and non-living elements. Originates from the French word 'environner.' Initially focused on physical aspects but now includes social, economic, and political spheres.
Abiotic Environment	Non-living elements that influence life on Earth. Includes physical factors like land (lithosphere), air (atmosphere), and water (hydrosphere).
Biotic Environment	Living organisms, including plants (flora), animals (fauna), and humans.
Lithosphere	The Earth's solid outer layer, consisting of the crust, mantle, and core.
Hydrosphere	All water on or near Earth's surface, including oceans, lakes, rivers, ice caps, and water vapor.
Atmosphere	The gaseous layer surrounding Earth, essential for life, extending roughly 300 kilometers above the surface.
Social Environment	Human-made features including economic, social, and political structures.

Component	Description
Economic Environment	Interactions between organisms and the physical environment to obtain resources for survival and growth.
Temperature	A fundamental ecological factor affecting the distribution and abundance of organisms. Varies with latitude and altitude.
Water	Essential for life, with availability varying greatly. Water quality and salinity also play crucial roles, especially in aquatic environments.
Light	A vital source of energy for autotrophs (organisms that produce their own food). Influences plant behavior and animal activities based on light intensity and duration (photoperiod).
Soil	Varies across the globe and is influenced by factors like climate, temperature, and humidity. Provides a foundation for terrestrial life.
Human Roles	Humans interact with the environment in threefold ways: physically (as biological organisms), socially (establishing institutions and policies), and economically (utilizing resources).

Facing Environmental Change: An Arsenal of Adaptations

Organisms inhabiting dynamic environments face a constant challenge: adapting to fluctuating abiotic conditions. This raises the question - how do they cope with these variations? The answer lies in a diverse toolkit of adaptations, allowing them to maintain a stable internal state (homeostasis) for optimal functioning.

Regulation: Maintaining the Balance

Some organisms, like birds and mammals, possess remarkable regulatory abilities. They can utilize physiological and sometimes behavioral mechanisms to maintain a constant body temperature (thermoregulation) and osmotic concentration (osmoregulation) regardless of external fluctuations. This internal stability is crucial for essential biochemical reactions and physiological functions. For instance, mammals like humans can sweat to cool down in hot environments and shiver to generate heat in cold conditions, all in a bid to maintain a core body temperature of around 37°C. Plants, on the other hand, lack such regulatory mechanisms and conform to external temperature changes.

Conforming to the Environment

The majority of animals (around 99%) and nearly all plants are conformers. Their internal environment passively reflects external conditions. For example, their body temperature fluctuates with ambient temperature, and aquatic organisms may experience changes in their internal osmotic concentration depending on the salinity of the surrounding water.

Migration: Seeking Refuge

Another strategy employed by some organisms is migration. This involves a temporary relocation to a more favorable environment during periods of stress. Similar to humans escaping the summer heat by moving to cooler locations, birds often undertake long-distance migrations to avoid harsh winters. A classic example is the arrival of thousands of migratory birds in the Keoladeo National Park, India, to escape the extreme cold of their Siberian habitats.

Suspending Activity: A Survival Tactic

For some organisms, surviving unfavorable conditions involves suspending activity. Bacteria, fungi, and lower plants form dormant structures like spores with thick walls, allowing them to wait out harsh conditions and germinate when the environment becomes suitable again. Higher plants utilize seeds and other vegetative structures to survive periods of stress and aid in dispersal. These structures can enter a state of dormancy, reducing metabolic activity and waiting for favorable conditions to germinate and establish new plants.

Similarly, animals that cannot migrate may resort to hibernation or aestivation. Hibernation, like that seen in bears, allows them to escape the harshness of winter by entering a state of dormancy with lowered metabolic rates. Aestivation, observed in some snails and fish, is a similar strategy employed to avoid extreme summer heat and desiccation. Additionally, zooplankton in some aquatic environments might enter diapause, a state of suspended development in response to recurring periods of unfavorable conditions.

The Living Stage: A Symphony of Organisms

The environment encompasses not just physical elements, but also the vibrant tapestry of life – the biotic components. This living stage is composed of interconnected populations from various species coexisting in a shared space. These populations can be broadly categorized into three main groups:

1. **Producers:** These self-sustaining organisms, also known as autotrophs (from Greek: auto - self, trophos - feeder), form the foundation of the biotic system. Primarily photosynthetic plants, they utilize sunlight to convert inorganic raw materials (like water and carbon dioxide) into complex organic compounds (food) through photosynthesis. In terrestrial ecosystems, rooted plants are the dominant producers. In aquatic environments, phytoplankton (floating microscopic plants) and macrophytes (shallow-water rooted plants) play this vital role.

2. **Consumers:** Heterotrophs (from Greek: heteros - other; trophs - feeder) are the consumers within the ecosystem. They rely on organic compounds produced by autotrophs for their sustenance. These consumers, also known as phagotrophs (phago - to swallow or ingest), can be further classified:
   1. **Herbivores:** These primary consumers, like cattle, deer, or grasshoppers, feed directly on plants. They represent the first link in the consumer chain.
   2. **Carnivores:** These predators feed on other animals. The hierarchy within carnivores is based on their feeding habits:
      i. **Primary carnivores** (second-order consumers) like foxes, frogs, or smaller fish prey on herbivores.
      ii. **Secondary carnivores** (third-order consumers) like wolves, owls, or peacocks feed on primary carnivores.
      iii. **Tertiary carnivores** (quaternary consumers) are at the top of the food chain, with lions and tigers as examples. These apex predators are not preyed upon by any other animal.
3. **Decomposers:** Also known as saprotrophs (from Greek: sapros - rotten; trophos - feeder) or reducers, these organisms play a critical role in breaking down dead organic matter (from plants and animals). Unlike herbivores and carnivores, decomposers don't ingest their food directly. Instead, they secrete digestive enzymes that break down the complex organic compounds in the dead matter, absorbing some of the decomposed products for nourishment. This decomposition process releases essential minerals back into the environment, which are then reused by producers, completing the nutrient cycle.

The Flow of Life: Mass and Energy Transfer Across Interfaces

Within an ecosystem, the movement of mass and energy across various boundaries is fundamental for its existence. These interfaces, where exchanges occur, play a critical role in maintaining the delicate balance of life. Here's a closer look at some key examples:

1. **Air-Water Interface:** This interface governs the exchange of gases like oxygen and carbon dioxide between the atmosphere and aquatic environments. Aquatic organisms rely on dissolved oxygen for respiration, while atmospheric oxygen is replenished by the photosynthetic activity of aquatic plants. Similarly, carbon dioxide, released by respiration and used by plants for photosynthesis, also undergoes exchange across this interface.

2. **Land-Atmosphere Interface:** Terrestrial ecosystems rely on this interface for gas exchange. Plants absorb carbon dioxide from the atmosphere for photosynthesis, releasing oxygen as a byproduct. Water vapor also transpires from plants into the atmosphere, influencing humidity and contributing to weather patterns.
3. **Soil-Plant Interface:** This crucial interface facilitates the uptake of essential nutrients and water by plants. Plant roots absorb dissolved minerals and water from the soil, transporting them throughout the plant for growth and development. Conversely, decaying plant matter and dead organisms return nutrients back to the soil, enriching it for future generations of plants.
4. **Decomposer-Organic Matter Interface:** Decomposers, like fungi and bacteria, break down complex organic matter (dead plants and animals) at this interface. Through the process of decomposition, they release essential minerals back into the soil, making them available for plant uptake. This recycling of nutrients is vital for ecosystem health.
5. **Organism-Environment Interface:** All living organisms interact with their environment across this interface. Animals acquire oxygen from the air or water for respiration and expel carbon dioxide as a waste product. They also take in nutrients and water from their surroundings for survival and growth.

Material Balance

The law of conservation of mass states that matter can neither be created nor destroyed, only transformed from one form to another. In a closed system (no mass transfer across boundaries), a material balance can be applied to track the quantity of a specific component.

Equation

Accumulation = Input - Output

Where:

1. Accumulation: The change in the mass of the component within the system over a specific time period. It can be positive (accumulation) or negative (depletion).
2. Input: The total mass of the component entering the system during the time period.
3. Output: The total mass of the component leaving the system during the time period.

Example

A continuously stirred tank reactor has a steady flow of water entering and exiting. To perform a material balance on water within the reactor, we would measure the mass flow rate of water entering (input) and exiting (output) over a specific time period. Any difference between these values would represent the accumulation of water within the reactor (positive) or its depletion (negative).

First Law of Thermodynamics

This law states that the total energy of an isolated system (no energy transfer across boundaries) remains constant. Energy can be transformed from one form (e.g., heat, work) to another, but the total amount is conserved.

Equation

$\Delta E = Q + W$

Where:

1. ΔE: Change in the total energy of the system over a specific time period.
2. Q: Heat transferred into the system (positive) or out of the system (negative).
3. W: Work done by the system on the surroundings (positive) or work done on the system by the surroundings (negative).

Example

A closed container filled with hot gas cools down as it loses heat to the surrounding environment (negative Q). The internal energy (E) of the gas decreases, and according to the first law, this decrease is balanced by an equal amount of heat transferred out (negative Q).

Second Law of Thermodynamics

This law introduces the concept of entropy, a measure of disorder or randomness within a system. It is denoted by "S". It states that in any isolated natural process, the entropy of the combined system and surroundings always increases over time (irreversible process). In other words, usable energy tends to become less available over time.

Equation

$\Delta S \text{ (total)} \geq 0$

Where:

4. **ΔS (total):** The change in total entropy of the system and the surroundings combined.

Example

As the hot gas in the previous example cools, its internal energy decreases, and the surrounding environment gains thermal energy (increases its entropy). The total entropy of the system and surroundings combined increases, reflecting the irreversible transfer of heat from a more ordered state (hot gas) to a less ordered state (warmer surroundings).

Heat Transfer: The Flow of Thermal Energy

Heat transfer is the fundamental process by which thermal energy, measured in kilojoules (kJ) and denoted by Q, moves between objects or systems due to a temperature difference. It's important to note that heat transfer only occurs when there's a temperature imbalance; there's no flow of heat between objects at the same temperature.

Heat transfer is the rate of thermal energy exchanged, symbolized by Q° (read as "Q dot"). This rate specifies how much heat is transferred per unit of time, similar to how speed describes the rate of change in position.

Heat transfer is directional, meaning it has a magnitude, direction, and specific point of action. A positive sign convention indicates heat transfer entering a system, increasing its internal energy. Conversely, a negative sign signifies heat transfer leaving a system, reducing its energy content.

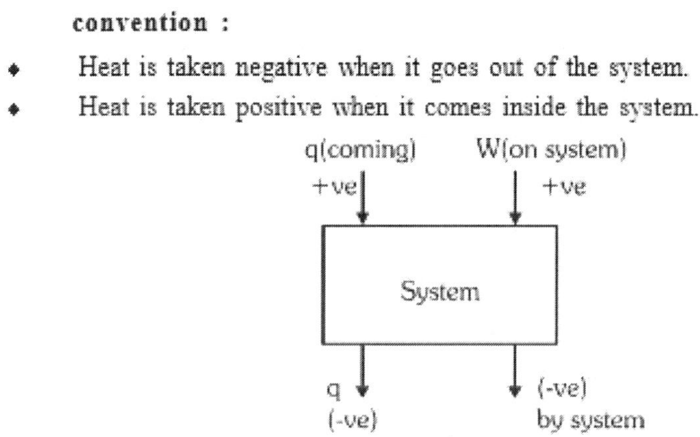

Figure 2: Heat Transfer system

There are three primary modes by which heat transfer occurs: conduction, convection, and radiation. All three mechanisms rely on a temperature difference to function.

i. **Conduction:** Imagine a hot cup of coffee. In conduction, heat transfer happens due to direct physical contact between objects at different temperatures. The more energetic particles (hot coffee) vibrate and collide with neighboring, less energetic particles (mug), transferring thermal energy through these interactions. This transfer of energy continues until thermal equilibrium is reached, meaning both objects reach the same temperature. Conduction is most effective in solids, where tightly packed particles readily transfer energy through vibrations and collisions.

ii. **Convection:** This mode involves the combined effects of conduction and fluid motion (liquids or gases). Consider heating a pot of soup. As the bottom layer of soup heats up from the burner (conduction), it expands and becomes less dense. This buoyancy effect causes the hot soup to rise, initiating a circular flow within the pot. The cooler liquid near the surface sinks, comes in contact with the hot burner (conduction again), and the cycle repeats. This continuous circulation of the fluid efficiently transfers heat throughout the pot. Convection can be natural, driven by buoyancy forces due to temperature variations, or forced, where external means like pumps or fans create fluid motion.

iii. **Radiation:** Unlike conduction and convection, radiation doesn't require any physical contact or medium for heat transfer. Instead, it involves the emission of electromagnetic waves (photons) from a hot object. These waves travel through space and can be absorbed by cooler objects, transferring thermal energy in the process. The Sun's heat reaching Earth is a prime example of heat transfer through radiation. Even hot objects in a vacuum can exchange heat through radiation.

Summary/Key Points

1. **Environment Definition and Origin:** The environment encompasses both living and non-living elements that influence an organism's existence. The term originates from the French word 'environner,' meaning to encircle or surround.

2. **Components of the Environment:** Natural components include land (lithosphere), air (atmosphere), and water (hydrosphere), along with diverse living communities. Human society and civilization have expanded the definition to include social, economic, and political spheres.

3. **Significance of Environmental Science:** Environmental studies help understand how human activities impact the environment and vice versa, fostering a balanced relationship between humans and nature. It involves various disciplines like physics, chemistry, biology, and sociology.
4. **Environmental Issues:** Global warming, ozone depletion, deforestation, resource depletion, and biodiversity loss are critical environmental challenges with far-reaching consequences for all life on Earth.
5. **Defining the Environment:** Various definitions highlight its comprehensive nature, encompassing both biotic and abiotic factors and the interactions between them.
6. **Classification of the Environment:** Environments can be classified into physical (abiotic), biotic (living organisms), and cultural (human-made) components, each with unique characteristics and interactions.
7. **The Physical Environment:** Divided into solid (lithosphere), liquid (hydrosphere), and gas (atmosphere) components, which shape habitats and influence the distribution of life.
8. **The Biotic Environment:** Comprises flora and fauna, with humans playing a significant role in shaping ecosystems and interactions.
9. **Interconnectedness of Life:** Organisms interact with their environment across various interfaces, exchanging mass and energy to maintain balance and sustain life.
10. **Heat Transfer and Thermodynamics:** Heat transfer occurs through conduction, convection, and radiation, guided by the laws of thermodynamics, which govern energy conservation and entropy increase.

Very Short Answer Type Questions

1. What is the environment?
2. What does the term "environner" mean?
3. Name three natural components of the environment.
4. Define abiotic environment.
5. Who are producers in an ecosystem?
6. What is the first law of thermodynamics?
7. Explain entropy in the context of the second law of thermodynamics.
8. How does convection differ from conduction in heat transfer?
9. Give an example of heat transfer through radiation.
10. What is the significance of studying environmental science?

Short Answer Type Questions

1. Define environment.
2. What is the origin of the term "environment"?
3. Name three natural components of the environment.
4. What is the significance of studying environmental science?
5. Define abiotic environment.
6. What are producers in an ecosystem?
7. Explain the first law of thermodynamics.
8. What is entropy in the context of the second law of thermodynamics?
9. How does convection differ from conduction in heat transfer?
10. Give an example of heat transfer through radiation.

Long Answer Type Questions

1. Discuss the expanding definition of the environment from its initial conception to encompassing social, economic, and political aspects.
2. Explain the importance of the environment for the survival of all organisms, highlighting key environmental issues.
3. Compare and contrast the definitions of environment provided by Douglas and Holland, C.C. Park, Encyclopedia Britannica, US Council on Environmental Quality, and K.R. Dikshit.
4. Describe the physical environment, including its divisions and characteristics.
5. Discuss the role of humans in the biotic environment, considering their physical, social, and economic aspects.
6. Explain the concept of material balance with an example.
7. Outline the three laws of thermodynamics and their significance in understanding energy and entropy.
8. Describe the process of heat transfer through conduction, convection, and radiation, including real-world examples.
9. Explore the interconnectedness of living organisms in an ecosystem, focusing on producers, consumers, and decomposers.
10. Discuss the flow of mass and energy across interfaces within ecosystems, highlighting the importance of these exchanges for ecosystem balance and stability.

Multiple Choice Questions

1. Environmental studies is a multidisciplinary science involving branches like:
   a. Biology only
   b. Physics only
   c. **Chemistry, physics, life science, and more**
   d. Mathematics only
2. The term "environment" originates from which language?
   a. Latin
   b. Greek
   c. French
   d. Sanskrit
3. Which of the following is NOT a component of the abiotic environment?
   a. Soil
   b. Plants
   c. Atmosphere
   d. Hydrosphere
4. What is the primary role of autotrophs in an ecosystem?
   a. Consuming organic matter
   b. Decomposing dead organisms
   c. Producing organic compounds
   d. Regulating temperature
5. The lithosphere is composed of:
   a. Solid components of Earth
   b. Gaseous components of Earth
   c. Liquid components of Earth
   d. All of the above
6. What law states that matter can neither be created nor destroyed, only transformed from one form to another?
   a. Second Law of Thermodynamics
   b. First Law of Thermodynamics
   c. Law of Conservation of Mass
   d. Law of Entropy
7. Heat transfer due to direct physical contact between objects at different temperatures is known as:
   a. Conduction
   b. Convection
   c. Radiation
   d. Conduction and convection
8. The environment can be broadly classified into which three categories?
   a. Physical, chemical, and biological
   b. Abiotic, biotic, and cultural
   c. Solid, liquid, and gas
   d. Producer, consumer, decomposer
9. What is the primary role of decomposers in an ecosystem?
   a. Producing oxygen
   b. Absorbing sunlight
   c. Breaking down dead organic matter
   d. Regulating temperature
10. What is the primary source of energy for photosynthesis in plants?
    a. Oxygen
    b. Carbon dioxide
    c. Water
    d. Sunlight

11. Which law states that in any isolated natural process, the entropy of the combined system and surroundings always increases over time?
    a. Law of Conservation of Mass
    b. First Law of Thermodynamics
    c. Second Law of Thermodynamics
    d. Law of Entropy
12. Heat transfer that involves the emission of electromagnetic waves from a hot object is known as:
    a. Conduction
    b. Convection
    c. Radiation
    d. Reflection
13. What are the components of the abiotic environment?
    a. Producers and consumers
    b. Living organisms
    c. Non-living elements
    d. Decomposers
14. Which environment consists of plants and animals?
    a. Physical environment
    b. Biotic environment
    c. Cultural environment
    d. Atmospheric environment
15. What are the modes of heat transfer?
    a. Conduction, convection, and reflection
    b. Conduction, convection, and radiation
    c. Conduction, refraction, and radiation
    d. Convection, radiation, and reflection
16. Which law states that the total energy of an isolated system remains constant?
    a. Second Law of Thermodynamics
    b. First Law of Thermodynamics
    c. Law of Conservation of Mass
    d. Law of Entropy
17. The process of breaking down dead organic matter by decomposers releases which essential nutrients back into the environment?
    a. Nitrogen
    b. Oxygen
    c. Carbon dioxide
    d. Minerals
18. Heat transfer due to fluid motion and temperature differences is known as:
    a. Conduction
    b. Convection
    c. Radiation
    d. Reflection
19. What is the primary role of heterotrophs in an ecosystem?
    a. Producing oxygen
    b. Consuming organic matter
    c. Absorbing sunlight
    d. Decomposing dead organic matter
20. What is the primary role of autotrophs in an ecosystem?
    a. Producing organic compounds
    b. Consuming organic matter
    c. Decomposing dead organic matter
    d. Regulating temperature

3

Fundamentals of Ecology

Introduction

The term "ecology" was initially employed by Ernest Haeckel in 1866. According to Haeckel, ecology is the scientific study of how living things interact with their surroundings. The word comes from the Greek word oikos, which means "home." The 'home life' of living things might thus be considered the subject of ecology (1). Ecology is the study of the properties of biological systems at various sizes, including species, populations, communities, ecosystems, and biomes. It is the methodical study of the relationships that exist between different living things and their environments. The environment is divided into two components: the physical or abiotic environment, which includes temperature, humidity, wind, soil pH, and minerals, and the biotic environment, which includes any influences on an organism exerted by other organisms, such as competition, predation, exploitation, and symbiosis (2-5).

Ecology has been of practical significance since the early stages of human history. In primitive societies, individuals required knowledge of their environment, including an understanding of natural forces and the surrounding flora and fauna, for survival. The onset of civilization coincided with the utilization of fire and various tools for environmental modification (6,7). Technological advancements have led to a diminished reliance on the natural environment for daily necessities. However, it is essential to recognize our ongoing dependence on nature for essential resources such as air, water, and indirectly, food, as well as for waste assimilation, recreation, and various other ecosystem services. Economic systems, regardless of political ideology, tend to prioritize human-made goods that primarily serve individual interests, while assigning minimal monetary value to natural goods and services that benefit society as a whole. Humans often take natural goods and services for granted until a crisis occurs, assuming they are unlimited or can be substituted by technological innovations. However, essential resources like oxygen and water, while recyclable, are not replaceable. While life support services are deemed free, they lack value within existing market systems (8-11).

Ecology encompasses three levels: the individual organism, the population (comprising individuals of the same species), and the community (comprising varying numbers of species populations). Ecology, at the organism level, examines the interactions between individuals and their environment, including the effects on and by these organisms. Ecology, at the population level, examines the presence or absence of specific species, their abundance or rarity, and the trends and fluctuations in their populations. Community ecology examines the composition and organization of ecological communities. Ecologists examine the pathways of energy and matter as they traverse between living and non-living components within the ecosystem, which includes both the community and its physical and chemical surroundings (12-14).

The interdependence of biotic and abiotic components is a fundamental principle of ecology. Biotic components are categorized into producers, which include plants and algae that engage in photosynthesis; consumers, encompassing herbivores, carnivores, and omnivores; and decomposers, such as bacteria and fungi that break down organic material. Abiotic components encompass sunlight, temperature, water, soil, and atmospheric gases, all of which significantly influence an organism's survival and behavior. The interaction between biotic and abiotic components of an ecosystem is complex and essential for maintaining ecological balance (15,16). The carbon, nitrogen, and phosphorus cycles must function continuously without depleting or destroying the essential components of the ecosystem. Energy enters ecosystems from solar sources, which producers convert into chemical energy through photosynthesis. This energy then flows through various trophic levels, represented by food chains and food webs. A key ecological principle pertains to ecosystem stability and resilienceEcosystems can recover from disturbances like fires, floods, or human activity as long as the core ecosystem's root structure and function are maintained. However, if an ecological threshold is crossed and an ecosystem transitions to a different stable state, it may lose biodiversity and ecosystem services (17,18). Since conservation biology and restoration ecology both aim to preserve and restore degraded ecosystems, it is critical to comprehend ecological instability and resilience. Ecology is also useful in a variety of fields, including forestry, urban planning, public health, and agriculture. Crop rotation, integrated pest management, and soil conservation are just a few of the sustainable agricultural techniques that are supported by the principles and expertise of applied ecology. The study of urban ecology focuses on creating cities that coexist peacefully with natural ecosystems, promote biodiversity and green infrastructure, and promote human health and well-being both within and outside of nature.

Organizational Levels of Ecological Systems

Multiple hierarchical levels of organization have been established to examine the interactions among organisms and between organisms and their physical environments. Ecological patterns and processes vary according to the organizational level at which they function. Ecologists have identified four fundamental levels of organization to analyze the interactions between organisms and their environment, as illustrated in Figure 5.

Abiotic and biotic components are present at every level of organization, creating a functional biological system defined by the exchange of matter and energy. Individual organisms, populations, communities, and ecosystems represent various levels of organization, which are interconnected. There are no distinct boundaries or breaks in function among these levels. A population is characterized as a group of organisms that coexist with other individuals of the same species. A biological community exists solely due to the food cycling and energy flux within the ecosystem. O'Neil et al. (1986) observed that energy dissipation in ecosystems could lead to the emergence of levels beyond those of individual organisms. Ecology encompasses various scales, ranging from the examination of individual organisms to the analysis of populations, communities, and ecosystems.

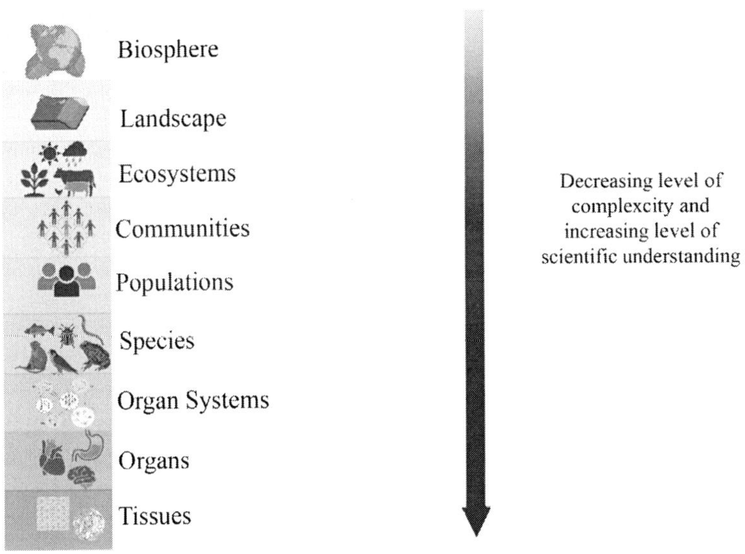

Figure 5: Organizational Levels of Ecological System

Individuals represent the most basic unit of ecological organization, encompassing a single plant, insect, or bird. Ecology examines the interactions between individual organisms and their environments, focusing on both the effects of the environment on organisms and the influence of organisms on their surroundings. Organismal ecology examines the behavior, physiology, morphology, and other characteristics of individual organisms in response to their environment. The population represents the subsequent level of biological organization. The phrase population possesses diverse definitions and implications across various academic areas. In ecology, a population refers to a collective of individuals of the same species inhabiting a specific geographical region. Population ecology examines the dynamics of populations, focusing on the mechanisms and reasons behind their temporal changes. Populations of various species within a specific area do not operate in isolation (25). They engage in communication with each other. Consequently, the community represents the subsequent, more advanced tier of organization within the interacting population of varied species. Ecological communities consist of interrelated populations of diverse species inhabiting a defined geographical region. Community ecology encompasses the composition, organization, and development of ecological communities. Communities exist on various scales, from small pom communities to extensive tropical rainforests. These communities represent, at their largest scale, a biome, a distinct natural assemblage of flora and fauna that dwell within a particular environment (for instance, tropical rainforests, coniferous forests, and savannas). It is characterized by unique vegetation that spans a vast geographical range, mostly influenced by local climatic circumstances. An ecosystem, or ecological system, is defined as the interacting assemblage of all living and non-living components within a physically delineated place. An ecosystem possesses boundaries as it is a system. An ecosystem is described as a system comprising interacting biotic and abiotic components. Ecosystems are self-organizing, self-regulating, complex, open systems with hierarchical organization. Ecosystem ecology examines the dynamics of energy flow and nutrient cycles within a community, as well as the interactions between species and their environment. It represents the pinnacle of an ideal ecology. It encompasses all of the Earth's ecosystems. The biosphere formally denotes all living species on Earth. In ecology, the biosphere, or ecosphere, is a functional concept that emphasizes the global interdependence of all living organisms and their environments. Several ecologists assert that 'Ecosystem' represents the fourth tier of organization for understanding ecological processes. Certain ecologists contend that the term "ecosystem" ought not to be utilized as the fourth level of organization, as it fails to encompass any novel biological organization. Some propose that the term "landscape" aptly describes a degree of organization characterized as an ecological system comprising multiple species.

Ecology examines the relationships and interactions among four tiers of biological order. The first level comprises organisms, defined as separate living entities. They react to their surroundings and engage with other creatures, exhibiting behaviors related to survival, reproduction, and competition. The second level comprises populations, which are assemblages of individuals of the same species residing in a certain geographic area. They engage with one another during mating, vie for resources, and occasionally assume social positions. The third level comprises communities, which are natural assemblages of diverse species populations inhabiting a single area and engaging in interactions shaped by predation, competitive exclusion, mutualism, parasitism, and other ecological dynamics. The fourth tier of ecological analysis is ecosystems, which comprise the populations of biotic communities and the abiotic elements of the environment (e.g., soil, water, climate, etc.). The four tiers of biological organization delineate how ecologists perceive and investigate the interactions between life and the environment.

Components and Structure of Ecosystems

An ecosystem is a dynamic combination of biotic and abiotic elements that work as a single, cohesive whole in the natural world. All living things, which fall into three main categories—producers, consumers, and decomposers—are considered biotic components. Green plants, algae, and a particular type of bacterium called cyanobacteria are examples of autotrophic creatures known as producers. They use photosynthesis to convert solar energy into chemical energy, which is the foundation of the food chain. As heterotrophic organisms, consumers get their energy indirectly by consuming other consumers or producers. According to their energy sources, they can be divided into three levels: primary consumers, which include herbivores like grasshoppers; secondary consumers, which are carnivores like mice that eat grasshoppers; and tertiary consumers, which are even higher level carnivores that get their energy from secondary consumers (31). Decomposers are the ultimate significant category of biotic components. Decomposers, including fungi and bacteria, are organisms that break down dead organic material. In doing so, decomposers are reintroducing nutrients into the ecosystem. The abiotic components encompass all living (biotic) organisms, as well as the non-living physical and chemical elements that influence and regulate ecosystem activities. The abiotic factors include sunshine, temperature, water, soil, air, minerals, and climate. Every abiotic component in the environment affects the survival, reproduction, and distribution of all creatures. The duration and intensity of light exposure on plant leaves influence the rate of photosynthesis and the amount of chemical energy transferred to primary consumers, such as grasshoppers. Temperature and water availability will functionally influence the metabolic activity of all organisms, ultimately influencing their dispersal. Collectively, these elements uphold ecological equilibrium, guarantee sustainability, and facilitate response to environmental alterations (32,33).

Ecosystem structure denotes the systematic organization of biotic (living) and abiotic (non-living) components within a system, emphasizing their spatial and functional relationships that facilitate ecological processes. The structure of ecosystems is fundamental to ecology as it provides insight into the composition, complexity, and functioning of natural systems (Table 3.1).

Table 3.1: Structural Components of Ecosystem

Components	Sub-category	Function	Example	References
Biotic	Autotrophs	Transform solar energy to chemical energy in photosynthesis; base of food webs	Cyanobacteria, algae, and green plants	34
	Heterotrophs	Consume other organisms for fuel; involved in energy flow and food web dynamics	Herbivores, carnivores, and omnivores like deer, lions, and humans, respectively	35
	Saprotrophs	Decompose dead organic matter and return nutrients to the system	Earthworms, fungi, bacteria	36
Abiotic	Soil	Provides nutrients and a medium for plant growth	Sand, clay, and loam soils	37, 38
	Light	Necessary for photosynthesis and regulating biological rhythms	Sunlight intensity and duration	39
	Water	Tied to physiological processes; shapes both aquatic and terrestrial life	Rivers, lakes, streams, rainfall	40
	Climate	Evaluates temperature, rainfall, and seasonal trends that impact organisms and processes	Wind, temperature, humidity, and precipitation	41, 42
	Air	Provides necessary gases (O_2, CO_2, N_2) to carry out respiration and photosynthesis	Atmosphere composition	43, 44
Tropic structure	Tropic levels	Classifies the way species are organized by energy source and efficiency of transfer of energy	1. Producers 2. Primary 3. Secondary 4. Tertiary	45
Spatial structure	Vertical Stratification	Specifies layers in ecosystems and provides light, humidity, and the niche of species	Forest floor, canopy, and understory	46
	Horizontal Distribution	Refers to the spatial pattern of organisms and habitats over landscapes	Patchy, or dumped distribution	47, 48

Trophic Pyramids and Food Webs

The trophic pyramid has served as a fundamental concept in ecology since its introduction nearly a century ago. The pyramid of numbers posits that the population of organisms at the base of food chains exceeds that of organisms at the apex of these chains. The concept was subsequently extended to the pyramid of biomass, illustrating that a substantial biomass of producers (plants) supports a smaller biomass of herbivores, which in turn sustains an even smaller biomass of carnivores, continuing up to apex predators. Utilizing thermodynamic principles, it is evident that each trophic level of the pyramid retains only a portion of the energy from the level beneath, resulting in the formation of the energy pyramid (49,50,51). Additionally, it was proposed that the slope of the pyramid, characterized by the increase in biomass across successive lower trophic levels, could indicate ecosystem stability. A survey of plankton types in the North Atlantic provided evidence that challenges the traditional bottom-heavy shape of trophic pyramids, indicating the presence of a 'inverted pyramid'. Comparable findings were observed in several other ecosystems. Conversely, another study suggested that this conflicting evidence indicates an overestimation of predator abundance or energy subsidies. A meta-analysis of 72 food webs concluded that inverted pyramids of species richness are expected to be rare.

It has recently been posited that inverted, top-heavy biomass pyramids can be stable and widespread in nature. Such pyramids can arise from several mechanisms, either endogenous or exogenous to the ecosystem. The initial mechanism pertains to transfer efficiency across trophic boundaries (e.g., biomass transfer exceeding 10%), which can be enhanced by improving the edibility and nutrient composition of food sources or by augmenting the foraging efficiency of consumers. A second significant factor is quick turnover rates, wherein herbivores reproduce more swiftly than their predators. Additional endogenous mechanisms encompass enhanced predator access to diverse prey pools, enabled by augmented resource output, and omnivorous predation as a strategy to consume both herbivores and producers, thereby promoting a rise in their biomass. Exogenous mechanisms of inverted pyramids involve the allocation of distant subsidies to predators and consumers, as well as the procurement of food from outside community boundaries to benefit local predators. For instance, top-heavy coral reef communities have been characterized by mobile consumers that derive energy from adjacent pelagic communities. Subsidies function across spatial and temporal dimensions, resulting in historical subsidies where elevated predator biomass may be supported by herbivores that were formerly part of the food web. Inverted pyramids have been observed in particular marine and estuarine environments, as well as in freshwater ecosystems such as streams, lakes, and ponds. Several examples have been documented in terrestrial ecosystems, albeit with significantly lower incidence. The validity of the fundamental concept of the trophic pyramid has not been tested across diverse terrestrial and marine animal classes simultaneously, especially on a global scale.

A food web represents the trophic interactions among organisms within a community, illustrating their dietary relationships. The analysis of food webs encompasses multiple branches, and these webs can be represented and illustrated in diverse manners to yield various inferences. The term food-web structure encompasses multiple properties. The most straightforward and prevalent application pertains to a graph illustrating species as nodes or vertices and their trophic interactions as links or edges with other species. The topology of this network constitutes the structure of the food web. This definition serves as our focal point, employing structure and architecture interchangeably with topology, while recognizing the existence of alternative interpretations of food-web structure, such as energy pathways, trophic pyramids, and food-chain lengths. Researchers aim to address a variety of issues that are nearly as diverse as ecology itself by examining food webs and their composition. In fact, it has been suggested that food webs offer a unifying framework that connects the traditional ecological subdisciplines (56,57). Energy flows and nutrient transfers, population dynamics, indirect effects, ecological niches and competitive interactions, species coexistence and biodiversity, specialization and coevolution, ecosystem functions, population and community stability, system resilience to environmental changes, and other phenomena are all examined by ecologists through the lens of food webs (58). Numerous studies discriminate between various types of food webs, including those that involve parasitoids and their insect hosts and those that are mostly composed of mutualistic relationships between plants and their seed dispersers, pollinators, and defenders.

Food web theory has progressed in two phases, beginning with qualitative and initial quantitative analyses in the 1970s and 1980s, focused on identifying universal structures and stability patterns within food webs. Significant research conducted between 1973 and 1980 initiated inquiries into fundamental aspects of trophic interaction dynamics, niche dimensions, and food-chain lengths (60,61). By the 1990s, empirical critiques identified the ecological shortcomings of overly simplified and taxonomically inconsistent “real” food webs, which frequently combined multiple taxa into broad categories, thereby undermining their reliability. Due to these issues, there were calls for higher resolution and the use of "trophic species" to standardize analyses. The second wave of food web theory emerged at the turn of the millennium, driven by advancements in computational power for the analysis of complex networks. The contemporary food web theory, known as network ecology, builds upon previous research by incorporating the study of food webs alongside mutualism, modularity, nestedness, and multilayer networks. Despite advancements in theory, current datasets remain a significant constraint for the application of network metrics. Current food-web datasets exhibit greater structural complexity and metadata richness compared to previous datasets. However, challenges persist due to various limitations, including taxonomic resolution errors, a scarcity of terrestrial networks, and instances of misidentified species and interactions. These issues hinder the reliable interpretation and application of network metrics, posing significant obstacles to advancements in ecological understanding.

Biogeochemical Cycles

Biogeochemical cycles enable the Earth's biosphere, atmosphere, hydrosphere, and lithosphere to recycle elements and compounds required for ecological balance, allowing life to prosper. Elements including carbon, nitrogen, phosphorus, sulfur, and water can travel through geological (rocks and soil), biological (living things), and chemical (reactions and processes) channels thanks to these cycles. For instance, plants take carbon dioxide from the atmosphere to start the carbon cycle. After being transmitted via respiration, food chains, photosynthesis, decomposition, or burning, that carbon dioxide eventually returns to the atmosphere (62,63). Scientists can comprehend and forecast the evolution of microbial communities and environmental activities by comprehending these cycles. Numerous activities can be used for good, such recovering precious metals like copper or uranium from low-grade ores or cleaning up organic and metal contaminants. Some negative parts of the cycles can lead to global environmental issues, such as the production of nitrous oxide, which can destroy the ozone layer, acid rain and acid mine drainage, and metal corrosion processes. As a result, understanding these cycles is becoming more and more important as the world's population grows and human activity has a greater impact on the environment.

Two biologically mediated chemical reactions are essential for understanding biogeochemical cycles: hydrolysis and oxidation-reduction processes. Hydrolytic reactions involve the chemical removal or addition of water to specific elements or molecules, typically accompanied by proton transfer, making them pH-dependent processes. Hydration and dehydration reactions thermodynamically store or release energy, influencing the reactivity and solubility of elements in the first series (H, Li, Na, etc.), as well as phosphorus, inorganic carbon, and silica in the environment. Redox reactions are defined by the transfer of electrons, the addition or removal of hydrogen atoms (excluding protons), or the involvement of molecular oxygen. Redox reactions serve as fundamental processes in biological systems, significantly influencing the distribution of oxygen, carbon, nitrogen, sulfur, and transition metals in the environment.

Carbon Cycle

Carbon is a crucial element for all life on Earth, constituting 50% of the dry weight of living organisms. Consequently, assessments of global production and the breakdown of organic carbon (OC) provide an approximation of the overall health index of the biosphere, both historically and currently. Life needs an energy transfer throughout the biosphere, predominantly driven by solar radiation, whereas the cycling of carbon through photosynthesis facilitates this energy flow. The carbon cycle resembles the energy flow on Earth, encompassing the metabolism of natural ecosystems, humans, and industrial systems. Plants convert solar energy into chemical energy within organic carbon molecules, which facilitate the biochemical processes needed for the evolution and utilization of environmental energy, a fundamental characteristic of life. It is continuously absorbed, released, and recycled through various natural and anthropogenic physical, biological, and chemical processes within the biosphere. It also generates an atmospheric greenhouse effect (similar to CO_2), regulates pH in seawater, and its redox properties facilitate several processes. The diverse components of carbon engage in biogeochemical cycling, a sequence of activities that distributes carbon among reservoirs and alters its molecular forms throughout these transfers. Consequently, global carbon cycling pertains to the exchange of carbon across various carbon reservoirs interconnected through a network of physical, chemical, and biological processes. The comprehensive carbon cycle comprises several interrelated cyclic pathways that vary for some reservoir processes. The principal reservoirs of carbon are: (i) the atmosphere, (ii) the oceans, (iii) terrestrial land (soil and vegetation), and (iv) the lithosphere.

Earth's crust contains carbonate rock, the greatest carbon reservoir. It is four orders of magnitude larger than the ocean's carbonate reservoir and six orders of magnitude greater than the atmosphere's carbon dioxide reservoir. The atmosphere, the smallest of the reservoirs, has the most carbon available for photosynthesis, which needs carbon dioxide. Since the late 19th century, when global industrialization started, human activity has impacted a number of the smaller carbon reserves. Deforestation and the use of fossil fuels have increased the atmospheric carbon dioxide reservoir while decreasing the quantity of fixed organic carbon in these reservoirs. Consequently, a net efflux of about $7X10^9$ metric tons of carbon dioxide per year has been released into the atmosphere. Over the past century, this imbalance has caused atmospheric carbon dioxide levels to rise by 28%, which has contributed to global warming through the greenhouse effect.

Over millions of years, atmospheric CO_2 is diminished via weathering processes involving silicate rocks and through the burial of fixed carbon by marine plants in marine sediments (67). Chemical weathering facilitates CO_2 sequestration as $CaCO_3$ in sedimentary rocks during warmer Earth temperatures, while releasing CO_2 into the atmosphere under colder conditions. This process has contributed to maintaining the Earth's climate within a narrow temperature range throughout geological history (68). Once disturbed, the Earth's climate will require hundreds of thousands of years to return to equilibrium through its natural carbon cycle processes. The primary forms of carbon in the atmosphere are carbon dioxide (CO_2) and methane (CH_4). These gases have significantly influenced the evolution of life forms and the transformation of Earth's surface environment over geological time. Carbon dioxide serves as the primary medium for photosynthesis, metabolic processes, and the decomposition of organic materials. The transformation in weathering and CO_3^{2-} precipitation contributes significantly to carbon cycling within the lithosphere. Carbon dioxide is the primary long-lived greenhouse gas that contributes to climate forcing. Since 1750, the radiative forcing has risen by 1.94 W m^{-2}, accounting for 65% of the total increase in forcing attributed to all long-lived greenhouse gases.

Nitrogen Cycle

Nitrogen makes up most of the Earth's atmosphere and is the fourth most abundant element in cellular biomass. The exchange between "reactive nitrogen" (those nitrogen molecules that support or are byproducts of cellular metabolism and growth) and inert dinitrogen gas (N_2) in the existing atmosphere is solely regulated by microbial activity (Figure 6). Since the planet's origin, nitrogen, an inert gas, has collected in the atmosphere and is one of the main planetary reservoirs. It is a limiting factor in primary production and is released during hydrothermal and volcanic eruptions.

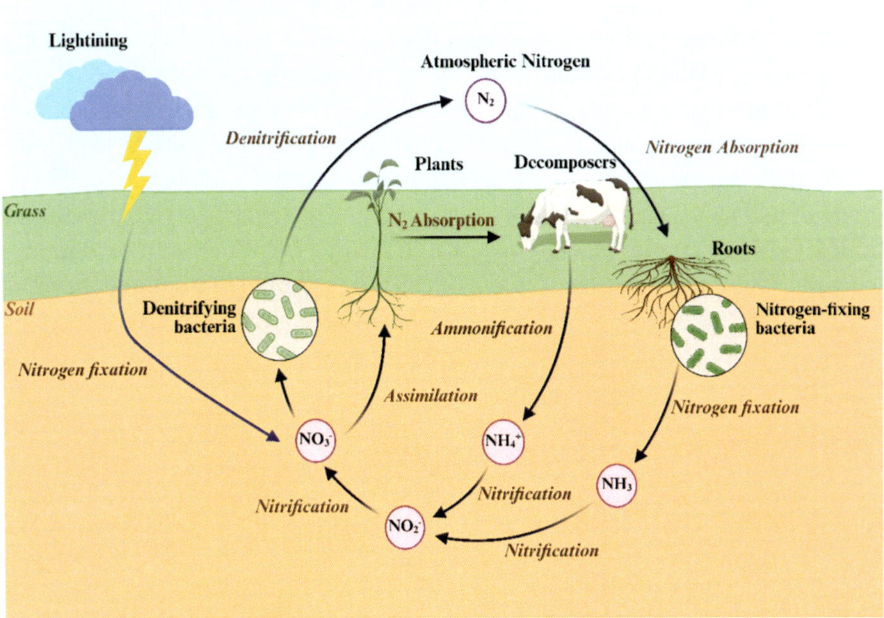

Figure 6: The Nitrogen Cycle

Nitrogen fixation is a slow and energy-demanding process performed by a restricted group of microorganisms, resulting in the availability of fixed nitrogen being a limiting factor. The nitrogen cycle is intricately linked to the carbon cycle, with fixed nitrogen distributed among various reservoirs, including organic nitrogen in living biomass, dead organic matter, and soluble inorganic nitrogen salts (71). Nitrogen is converted into ammonia by more than 100 free-living bacteria, which encompass both aerobic and anaerobic bacteria, actinomycetes, and cyanobacteria. These organisms inhabit various environmental niches and fix nitrogen under specific conditions. Cyanobacteria serve as the primary nitrogen-fixing organisms in aquatic ecosystems, exhibiting elevated fixation rates attributable to their photosynthetic capabilities. Plants and microbes enhance N_2 fixation efficiency through symbiotic relationships, exemplified by the Rhizobium-legume association, which can elevate fixation rates to 200-300 kg N/hectare/year. Nitrogen fixation is restricted to bacteria and requires significant energy, thus ammonia and oxygen can regulate the process.

In predominantly oxidized environments, fixed inorganic nitrogen predominantly exists as nitrate (NO_3) or, to a lesser degree, nitrite (NO_2). The two oxidized forms of nitrogen are produced from ammonium through the activity of nitrifying bacteria. Nitrification is a metabolic process that involves the oxidation of ammonium, which can be coupled with the reduction of inorganic carbon to synthesize organic molecules in the absence of light energy. The metabolic process referred to is known as chemoautotrophy. The process of ammonium oxidation is divided into two categories: organisms that oxidize ammonium to nitrite and those that subsequently oxidize nitrite to nitrate. Nitrification is a metabolically inefficient process, as the oxidation of one mole of NH3 yields only one mole of organic carbon. Consequently, the growth rate of nitrifying bacteria is typically low, while the rate of ammonium oxidation is generally rapid in the presence of O_2. Nitrification occurs in a limited group of eubacteria. Gene sequence analysis indicates that the pathways have evolved from a single ancestral gene and exhibit a high degree of conservation. Nitrate (NO_3) is a thermodynamically stable molecule in the oxidizing conditions of the modern environment, making it the most common form of fixed nitrogen in the world's oceans. The nitrogen cycle concludes with the reduction of oxidized nitrogen species, particularly nitrate (NO_3), resulting in the formation of nitrous oxide (N_2O) and dinitrogen (N_2). Denitrification is an anaerobic process facilitated by a variety of bacteria and Archaea. Denitrification is a respiratory process that utilizes nitrate (NO_3) or nitrite (NO_2) as the terminal electron acceptor. Organisms opportunistically reduce nitrate to nitrous oxide and nitrogen dioxide, leading to a considerable loss of fixed nitrogen in the environment. This reduction occurs mainly in anaerobic sediments within marine and lacustrine environments, as well as in organic carbon-rich soils that experience periodic flooding. Fifty percent of nitrogen fixation on Earth results from natural biological processes, whereas the remaining fifty percent is attributed to intentional fixation through human activities. The Haber reaction, established in the 19th century, serves as the principal source of nitrogen for fertilizers in commercial agriculture. Approximately 25% of applied nitrogen is lost via denitrification, resulting in an increase in atmospheric N_2O levels. A significant portion of this fixed nitrogen seldom reaches open oceans.

Sulfur Cycle

The sulfur cycle involves the conversion of sulfur and other nutrients into sulfate by soil and rhizosphere-inhabiting microorganisms. This cycle comprises three stages: the transition of sulfur between sulfide deposits and sulfate, the exchange between soil biomass reservoirs and soluble SO_4^{-2}, and the interaction between complex sulfide and SO_4^{-2} in soil with forms accessible to plants. Soil temperature, water content, reaction rate, and microorganism activity in the rhizosphere influence these transformation processes. This cycle can be classified into five reactions: mineralization, immobilization, oxidation, reduction, and volatilization. Sulfur comprises only 1-3% of microbial biomass and 0.1-1% of dried weight. Carbon, Nitrogen, and Sulfur cycling are significantly influenced by α-proteobacteria.

The breakdown of organic sulfur compounds into inorganic forms by microbial activity is known as sulfur mineralization. These compounds can either be released into the environment as oxides or used by plants or microbes for the production of cell components. Numerous inorganic and organic sulfur compounds are stored in the cells of bacteria and fungi and are only released when the cells are lysed. Protein putrefaction under waterlogged anaerobic conditions causes hydrogen sulfide (H_2S) and odoriferous mercaptans to accumulate in soils. Estimates for annual sulfur mineralization rates range from 1.0 to 10.0%, and they are strongly correlated with nitrogen mineralization rates. Heterotrophic microorganisms use energy from organic carbon to develop and multiply, whereas autotrophic microbes directly mineralize sulfur into sulfides. The chemical oxidation of additional cysteine to cystine in soils, followed by the oxidation of cysteine to sulfate with Cys-disulfoxide and cysteine sulfonic acid as intermediates, is the mechanism by which sulphate is formed. The primary enzymes that cause sulfur-esters to mineralize are rhodanese and sulfatases. Immobilization is the process by which sulfur compounds are assimilated into organic forms in microbial and plant cells through covalent bonds, rendering them even less accessible to microorganisms and plants. creatures consume plants and other creatures to meet their sulfur needs. Protein amino acids, sulfur-esters, sulphonates, vitamins, taurine, and co-factors are the main forms of sulfur accumulation (82). Numerous crop plants produce sulfur-containing organic compounds, such as glucosinolates, which provide defense against herbivores, fungus, insects, and microbes. Depending on the plant species, agricultural leftovers containing sulfur metabolites decompose and mineralize at different speeds before being immobilized by the plants or the microbial community (83,84).

Bacterial chemosynthesis and photosynthesis involve the process of sulfur oxidation, in which certain sulfur-oxidizing microorganisms oxidize S°, H2S, and sulfide (S^{2-}) to create SO_4. This procedure increases electrical conductivity, restores alkaline soils, and increases soil fertility. Pond bottoms turn black as a result of sulfide oxidation, a significant transformation process in ponds or stagnant water systems. Because reduced sulfur compounds are unavailable in sulfur-deficient agricultural soils, heterotrophic sulfur-oxidizing bacteria are essential. The population of autotrophic sulfur-oxidizing *Thiobacillus* species has been shown to significantly increase following the application of elemental sulfur in soils. The addition of organic substrates to soils may increase the number of heterotrophic microbial communities, resulting in better sulfur oxidation. Bacteria that oxidize sulfur can be categorized into three types: chemoautotrophic sulfur-oxidizing bacteria, which are obligate aerobes capable of catalyzing a series of sulfur compounds; chemolithotrophic sulfur-oxidizing bacteria, which can utilize ferrous iron as an energy source and produce thiosulfate in addition to sulfur; and finally, tetrathionate-producing sulfur-oxidizing bacteria. There are two strategies to reduce and volatilize sulfur compounds: (i) dissimilatory sulphate reduction and (ii) assimilatory sulphate reduction. By employing sulfate as a terminal electron (e^-) acceptor, the latter pathway releases H_2S. There are 220 species of sulfate-reducing bacteria in 60 genera that have been identified. Some of the most common bacteria that reduce sulphate include *Desulfobulbus, Desulfobacter, Desulfococcus*, and *Desulfosarcina*.

The reduction of sulfur or sulfate is detrimental as it diminishes sulfate availability, soil fertility, and crop productivity. It contributes to the pollution of canal water, which may be utilized for crop irrigation. Sulfate-reducing bacteria play a role in the biogenesis of sulfides of cobalt, iron, antimony, cadmium, lead, zinc, and nickel, influenced by the toxicity of metal ions in specific environments. Moreover, the emission of inorganic or organic sulfur gases in the environment constitutes a significant transformation process within the sulfur cycle. Numerous sulfur gases are emitted from anthropogenic sources, including the combustion of fossil fuels, as well as from soils, marshes, sediments, and peats. Various volatile sulfur compounds, including H_2S, methyl mercaptans (CH_3SH), dimethyl sulfide (CH_3SCH_3), carbonyl sulfide (COS), dimethyl disulfide (CH_3SSCH_3), and carbon disulfide (CS_2), are produced by different fungi and heterotrophic bacteria during the metabolism of sulfur compounds.

Phosphrous Cycle

The phosphorus cycle is a vital biogeochemical cycle that describes the circulation of phosphorus through the lithosphere, hydrosphere, and biosphere. Unlike carbon, nitrogen, or sulfur cycles, phosphorus does not have a significant gaseous phase and mainly exists in solid and aqueous forms. Its largest reservoir is in rocks and sediments, from where it gradually enters ecosystems. Phosphorus is an essential nutrient because it is a key component of DNA, RNA, ATP, and phospholipids, all of which are necessary for energy transfer, cell function, and the structural integrity of living organisms.

The cycle begins with the weathering of phosphate-rich rocks, which releases phosphate ions into the soil and water. These phosphate ions are absorbed by plants through their roots and incorporated into organic molecules such as nucleic acids and energy compounds. When animals feed on plants, phosphorus moves through the food chain. It is further transferred to carnivores when they consume herbivores. Thus, phosphorus flows from one trophic level to another, supporting life processes across ecosystems.

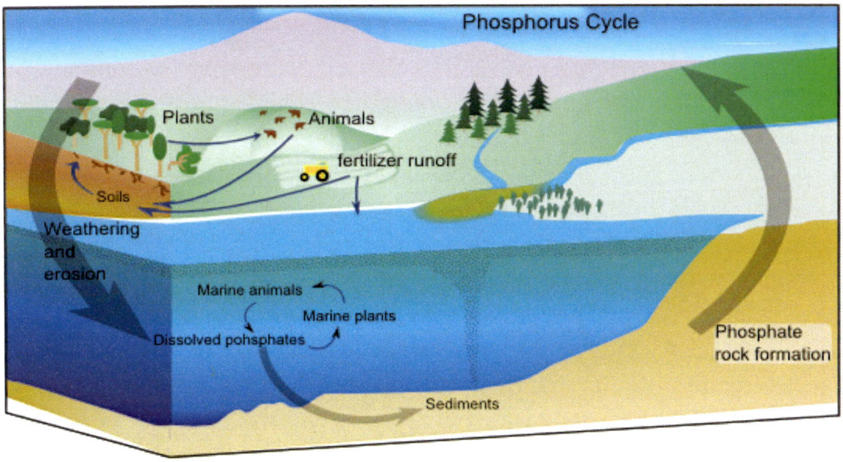

Figure 7: Phosphorus Cycle

When plants and animals die, or when animals excrete waste, phosphorus returns to the soil or water in the form of inorganic phosphate. Decomposers such as fungi and bacteria play an important role by mineralizing organic phosphorus into inorganic forms that can be reused by plants. Some phosphorus is washed away by rain or irrigation and carried into rivers, lakes, and oceans. In aquatic systems, dissolved phosphates may settle on the seabed, eventually becoming sediments. Over geological time scales, these sediments may undergo uplift to form new phosphate rocks, restarting the cycle.

The phosphorus cycle is considered a sedimentary cycle because its primary reservoir is in rocks and sediments, and it is often much slower than other nutrient cycles. It is also a limiting nutrient in many ecosystems, especially freshwater environments, where low phosphorus availability restricts plant and algal growth. Human activities have greatly disrupted the phosphorus cycle. Mining phosphate rocks for fertilizers and detergents has increased phosphorus input into ecosystems. Excessive use of fertilizers leads to runoff into water bodies, causing eutrophication, algal blooms, oxygen depletion, and massive fish kills. Deforestation and soil erosion also accelerate phosphorus loss from terrestrial ecosystems. diagram enclosed

Population Dynamics and Growth Patterns

Populations of organisms can be examined as complicated feedback systems with both positive and negative feedback loops. The numbers of creatures alter in fascinating and strange ways. Some, such as the human population, appear to be growing at a constant rate, but others, such as the blue whale, appear to be extinction-prone. Growth patterns like these have significant ramifications for the species in question, people who use them as resources, and the survival of life on Earth. Other species, such as the sycamore aphid in England, tend to be quite stable over long periods of time, despite significant fluctuations from year to year. Other populations, such as the larch budmoth in the Swiss Alps, experience spectacular and very regular multi-annual oscillations that can take up to ten years to repeat themselves. Ecologists have been captivated by these so-called "population cycles" for over 50 years. There is still no universal consensus on the causes. Furthermore, species exhibit far more varied dynamics. Gypsy moth egg mass counts in New England, for example, declined suddenly and, at the same time, switched from a saw-toothed to a cyclical pattern of fluctuation, while whitefish in Lake Ontario remained relatively constant for a long time but recently entered a period of decline.

Changes in a dynamic system can be induced by either external or internal processes. For example, the predator population has an exogenous effect on the prey population since it causes changes in prey numbers while being unaffected by the changes. Endogenous effects, on the other hand, cause changes in a dynamic variable and are influenced in turn by those changes (93,94). Feedback loops form when one variable influence another, either directly or indirectly. Dynamic systems are deemed stable when their variables recover to or near their original states after disruptions. A room's temperature is stable when the thermostatic control system returns it to its former temperature, or an equilibrium point. Stable systems are homeostatic or controlled, and their equilibrium point may never be observed if subjected to constant exogenous disruption. Stability in dynamic systems is achieved by negative feedback processes, which oppose changes in a dynamic variable. The thermostat is a mechanical feedback device, but natural ecosystems can also generate feedback mechanisms, such as the second-order feedback loop caused by population interaction. Exogenous elements like as temperature, rainfall, soil type, and topography create the conditions for dynamic interactions. For example, gradual temperature rises caused by global warming might alter the interaction structure of biological systems. Understanding how feedback loops are formed in ecological systems is critical for producing stable, self-sustaining ecosystems.

Species Interactions and Community Characteristics

Species interactions play a crucial role in ecological systems, forming communities and impacting ecosystem functions. The fundamental components of ecological networks, which can swiftly expand in complexity beyond a few basic paired relationships, are interactions such as mutualism, predation, parasitism, commensalism, and competition. In actuality, species interact far more frequently in higher-order interactions, where the interaction between two species is dependent on the presence or absence of one or more other species. This is in contrast to ecological models, which usually assume pairwise interactions between species in communities. A microbial species may, for instance, produce an antibiotic that stops a competitor's growth, and a third microbial species may produce an enzyme that breaks down the antibiotic. This would ultimately change the initial inhibitory interaction on the first competitor, but crucially, neither of the species operating alone would be impacted. A fourth species might likewise inhibit the enzyme activity, resulting in a four-way interaction. These kinds of cascading effects offer plenty of opportunities to illustrate how species affect one another indirectly through intricate chains of interactions as well as directly. Higher-order interactions can result from adaptive reactions, such as when prey reduce their foraging behavior in response to predators or when predators change the sorts of prey based on the availability of food. In such instances, adaptive reactions may have further ramifications for ecological relationships: a third species may become more active in foraging as a result of less competition, which may ultimately impact a fourth species. Thus, standard pairwise-based conceptualizations are unable to adequately capture the intricate food webs of trait-mediated and density-mediated interactions created by indirect effects. As a result, ecological networks are contextual and dynamic, changing over time due to component-level variability and species features. Understanding ecosystem resilience and forecasting the successive effects of perturbations, including species loss and introduction, need on new techniques for ecological network analysis that quantify network properties like modularity, nested Ness, and connectome. To effectively simulate and protect ecological systems in a world impacted by global environmental change, an integrative and multifaceted perspective of species interactions is necessary.

In ecological systems, community characteristics are combinations of structural and temporal features that define how species cohabit, interact, and evolve throughout time. The temporal dynamic and how communities evolve in terms of species richness or composition over time are among these traits' most crucial features. Two characteristics can be used to assess these temporal changes: the characteristic Jaccard index, which characterizes the degree of compositional turnover or similarity in time-lags, and the community characteristic time, which is a time-scale of species' (species richness) observed changes. These indices make it possible to determine if a community is stable or changing quickly, which reflects resilience and how an ecosystem or community persists after an environmental disturbance. Although characteristic time and Jaccard indices offer comparable insights into temporal change, communities with short characteristic times will typically undergo more frequent changes, indicating a high rate of species turnover and shorter periods of species residents, in contrast to communities with long characteristic times, which typically have more stable and persistent periods of species residents. Frequent sampling in communities that are changing quickly, for instance, would yield richer sample data and analysis on community viewpoints. This association between sampling frequency and characteristic times is also significant. The intricacies of community ecology can be thought of via the simple lens of comparing species changes across time. For long-term monitoring, conservation planning, climate change responses, and other deductions regarding anthropogenic impacts, this approach is particularly crucial (98).

Conclusion

Ecosystems encompass a range of systems, from individual organisms to the entire Earth. Each system consists of complex interactions between biotic and abiotic components. The structure and function of ecosystems are sustained by energy transfer across trophic levels and food webs, alongside nutrient cycling through biogeochemical processes such as the carbon and nitrogen cycles. Furthermore, population growth and environmental changes contribute to alterations in species distribution across temporal and spatial dimensions. Interactions among species within a community influence the composition of an ecosystem, and the regions that define these interactions will determine the stability and resilience of the ecosystem. Comprehending these interrelated concepts is essential for predicting ecological responses to disturbances and facilitating sustainable environmental management and biodiversity conservation.

Summary

Ecosystems range from individual organisms to the entire Earth. They consist of biotic (living) and abiotic (non-living) components interacting together. Energy flow through trophic levels and food webs sustains ecosystem functions. Nutrient cycling (e.g., carbon and nitrogen cycles) maintains balance. Population growth and environmental changes affect species distribution over time and space. Species interactions shape community composition and ecosystem dynamics. The stability and resilience of ecosystems depend on these interactions.Understanding these interconnections is key for predicting ecological responses and ensuring sustainable management and biodiversity conservation.

Short Answer Type Questions

1. What is ecology?
2. What is a population?
3. Define community.
4 Give an example of an abiotic factor.
5. What is a food chain?
6. What are producers?
7. Name one primary consumer.
8. What is biodiversity?
9. What is the ultimate source of energy for ecosystems?
10. Which organisms are called decomposers?

Short Answer type Questions

1. Differentiate between food chain and food web..
2. What are abiotic and biotic components of an ecosystem?
3. Explain ecological niche.
4. What is ecological succession?
5. Distinguish between autotrophs and heterotrophs.
6. Give an example of mutualism.
7. What is the difference between a population and a community?
8. Define the term trophic level.
9. Why are decomposers important in nature?
10. Explain the 10% law of energy transfer

Long Answer Type Questions

1. Describe the structure and components of an ecosystem.
2. Explain energy flow in an ecosystem with a diagram.
3. What is ecological succession? Explain types with examples.
4. Discuss biotic interactions with examples.
5. Explain food chains and food webs with examples.
6. What is biodiversity? Explain its importance.
7. Describe population interactions in detail.
8. Explain the nitrogen cycle in nature.
9. Write a detailed note on ecological pyramids.
10. Discuss the threats to biodiversity and conservation methods

Multiple Choice Questions

1. Which of the following is the ultimate source of energy for all ecosystems?
   a) Moon **b) Sun**
   c) Earth's core d) Wind
2. The pyramid of energy in an ecosystem is always:
   a) Inverted **b) Upright**
   c) Irregular d) Cylindrical
3. Primary consumers in a grassland ecosystem are:
   a) Lions **b) Deer**
   c) Tigers d) Vultures
4. Which organisms are called 'nature's scavengers'?
   a) Carnivores b) Herbivores
   c) Decomposers d) Omnivores
5. Ecological niche refers to:
   a) Physical space occupied by an organismb) Functional role of an organism in ecosystem
   c) Both a and b d) None of these
6. Which process converts ammonia into nitrates?
   a) Nitrogen fixation **b) Nitrification**
   c) Denitrification d) Ammonification
7. The interconnected network of food chains is known as:
   a) Food niche b) Food cycle
   c) Food web d) Food chain
8. In ecological succession on bare rock, which organisms appear first?
   a) Shrubs **b) Lichens**
   c) Mosses d) Herbs
9. Which pyramid can be inverted in an aquatic ecosystem?
   a) Pyramid of energy b) Pyramid of numbers
   c) Pyramid of biomass d) None of these
10. 10% law of energy transfer was given by:
    a) Tansley b) Elton
    c) Lindeman d) Odum

Section II
Biodiversity and Natural Resources

4

Biodiversity and Environment

Have you ever wondered why you and your siblings appear different, despite sharing the same biological parents? Or why nature boasts such a diverse array of plant and animal life? Or what purpose the numerous species in nature serve? If these questions intrigue you, this chapter will provide the answers you seek. The various forms of living organisms found in nature are collectively known as **biodiversity**. This term, which was introduced by E.O. Wilson in 1985, describes the richness and variability of species among plants, animals, microorganisms, and other life forms. The word **"biodiversity"** itself conveys its meaning: the diverse range of living organisms inhabiting Earth. It encompasses the variety and variability among different types of living creatures, including flora, fauna, microscopic organisms, and aquatic life. Biodiversity is crucial for life on Earth, as these species and organisms interact to sustain one another. It plays a vital role in our survival by providing a wide array of organisms that work together to supply food, water, shelter, medicine, and other necessities essential for survival of living beings.

Types of Biodiversity

Biodiversity can be classified into three primary categories:

1. Species diversity
2. Genetic Diversity
3. Ecological Diversity

1. **Species diversity**, which occurs when various species coexist in a specific area. e.g.- coral reefs, rainforests, freshwater ecosystems etc.
2. **Genetic diversity**, referring to the differences in genetic makeup among individuals of the same species, resulting in unique appearances. E.g.- varieties of flora and fauna, humans etc.
3. **Ecological diversity**, which encompasses the range of ecosystems within a particular geographic region. An ecosystem is a functional unit comprising living organisms (biotic elements) and their physical environment (abiotic elements). These ecosystems can be classified

as terrestrial, freshwater, marine, or artificial. Ecological diversity emphasizes the distinctions and variations among these different ecological systems within a given area. e.g.- forests, grasslands, tropical rainforests, estuaries etc.

Species richness and evenness

Species richness and evenness are two important concepts in biodiversity-

Species richness refers to the wide array of species present in an ecosystem. For instance, tropical rainforests are known for their high species richness, hosting numerous plant species, but may have low species evenness, with fewer plants of certain species. Conversely, **species evenness** describes a situation where members of one species are plentiful, but there is less variety in the types of species present. As an example, a researcher might discover a large population of fish belonging to the same species in a pond (high species evenness), but with limited diversity (low species richness).

Megadiversity zones and hotspots

Species-rich areas known as biodiverse zones harbor a wide array of plant and animal life, including vertebrates and invertebrates. Many of these species are unique to their specific regions and cannot be found elsewhere on the planet. The distribution of biodiversity across the Earth is uneven, with 17 key megadiversity countries occupying just 10% of the Earth's surface but containing a vast array of the world's species. These megadiverse areas cover only 24.46% of the Earth's land mass. According to Conservation International, these regions encompass over 70% of the planet's biodiversity. Among these megadiversity nations are India, the United States, Brazil, Indonesia, China, Colombia, Mexico, Australia, Peru, and Venezuela etc. These 17 countries collectively host a diverse range of flora, fauna, vertebrates, and invertebrates, supported by favorable soil conditions and climates that promote species survival. These nations are spread across four continents, which will be explored below.

American Continent

The first and one of the largest continents includes seven countries: Brazil, Colombia, Ecuador, Mexico, Peru, the United States of America, and Venezuela.

1. **Brazil**, occupying half of South America, is home to the Amazon rainforest, contributing to its rich biodiversity. Some of its endemic species include the *Apoclada* bamboo plant, the Ashy Black Titli monkey,

Plecturocebus cinerascens and the Brazil-nut poison frog *Adelphobates castaneoticus*. Brazil contains 20% of the world's freshwater and boasts 43,893 plant species, 1,900 bird species, over 978 amphibian species, and more than 700 species of reptiles and mammals.

2. **Colombia**, situated between South and Central America, is recognized for having the second-highest number of plant and animal species. It features diverse landscapes and invests significantly in conservation efforts. Colombia's biodiversity includes over 1,800 bird species, with 73 being endemic, such as the Chestnut-winged Chachalaca and the Sooty-capped Puffbird.
3. **Ecuador**, the smallest country among the 17 megadiversity nations, is located on the fringes of the Galápagos Islands. This highly diverse country is home to numerous volcanoes and lakes. It hosts more than 16,000 plant species, including 4,173 endemic species. The island is also home to several endemic species of amphibians, mammals, and over 4,000 plant species. Some of Ecuador's native species include the Ecuadorian Sac-winged Bat, the El Oro Parakeet, and the Tanti Rocket Frog.
4. **Mexico**, considered the world's second-largest ecosystem, is home to approximately 200,000 distinct species. The country has 536 mammal species, with 30% being endemic. It also hosts numerous endemic reptile and amphibian species. Mexico is renowned for its unique flora, fauna, aquatic life, and cenotes. It boasts over 25,000 plant species. Some of the endemic wildlife found in Mexico includes the Chinanteco Deer Mouse, the Cozumel Thrasher, and the Alamos Mud Turtle.
5. **Peru** boasts diverse geographical features, including the Amazon rainforest, Andes mountains, and Pacific coastlines. Forests cover over 58% of its land, hosting more than 30,000 plant species and iconic animals like tapirs and jaguars. The country is home to 70 endemic mammal species, such as the Yellow-tailed Woolly Monkey.
6. **The United States**, the world's largest nation, significantly contributes to biodiversity conservation and enhances its natural beauty. It harbors numerous endemic species, with over 90% of its wildlife being unique to the region. The country supports more than 17,000 plant species, including endemic creatures like the Channel Islands Spotted Skunk on Hawaii.
7. **Venezuela** hosts 21,000 plant species throughout its territory, along with 315 amphibian, 341 reptile, 351 mammal, and over 1,000 bird species.

Approximately 38% of its plant species are endemic. Notable endemic animals include the Fiery Squirrel, Sira Poison Frog, and Sucre Spiny Rat.

Asian Continent

Asia encompasses five megadiverse countries: China, the Philippines, Indonesia, India, and Malaysia.

1. **India,** one of the world's most biodiverse nations, is home to about 1,250 bird, 410 mammal, 408 reptile, and 197 amphibian species. It features four megadiverse zones: the Himalayas, Nicobar Islands, North-East India, and Western Ghats. Endemic species include the Black-footed Gray Langur, Jungle-striped Squirrel, Malabar Civet, and Nicobar Tree Shrew. Biodiversity conservation is crucial for sustaining essential human resources. India also shelters endangered species like the wild yak, Tibetan argali, snow leopard, and Asiatic ibex.
2. **Indonesia's** varied climate fosters exceptional biodiversity. It houses approximately 1,592 bird species, 515 reptiles, and 270 amphibians. The country boasts over 25,000 floral species, with 55% being endemic.
3. **China**, another vast and diverse nation, is particularly rich in avian life, hosting 1,244 bird species. It is also home to 667 endemic vertebrate species and roughly 30,000 plant species. Despite facing challenges from land clearance and pollution, China maintains its biodiversity.
4. **The Philippines**, though densely populated, supports 612 bird species (including 194 endemic ones), 270 reptile species, and 111 amphibian species. It showcases floral diversity with over 9,250 native plants, 33% of which are endemic. Malaysia is renowned for its lush flora, mountains, tropical forests, and coral reefs.
5. **Malaysia-** The equatorial climate of Malaysia, characterized by high humidity, contributes to its abundant biodiversity. However, ecosystems and certain species in the country, particularly the Bornean orangutan, are under considerable pressure from deforestation, especially in Borneo Island's Sarawak and Sabah regions. As one of the world's megadiverse nations, Malaysia hosts 210 mammal species, 620 bird species, 250 reptile species, and 150 frog species. The nation is home to distinctive native fauna, including Leonard's Pipe Snake, Hose's Pygmy Flying Squirrel, and the eye-catching Woolly-Stalked Begonia. Its extensive rainforest cover and lengthy coastline provide habitats for diverse wildlife. The vast forested areas also support the growth of 14,500 plant species throughout the country.

African Continent

1. **South Africa's Cape Floral Region** features a variety of unique plant species. The nation is renowned for its 'Big Five' wildlife: elephants, lions, leopards, buffaloes, and rhinoceroses. Recognized as a megadiverse country, South Africa is home to 858 bird species and 299 mammal species. Of the 135 frog species found in the nation, 110 are endemic. Several animals are exclusive to southern Africa, such as the long-nosed Golden Mole and the Riverine Rabbit. The country also harbors other notable fauna, including giraffes, cheetahs, and hippopotamuses.
2. Endemism refers to species unique to a specific geographic area and not found elsewhere. **Madagascar** exemplifies this, with about 90% of its flora and fauna being endemic. Notable examples include lemurs and baobabs, with 6 out of 8 baobab species native solely to Madagascar. The island's diverse ecosystems place it among megadiverse countries. It hosts over 13,700 plant species (mostly endemic), 389 reptile species, 278 amphibian species (all endemic), 282 bird species, and 220 mammal species. Exclusive animals include the carnivorous Fossa, Cuckoo-roller bird, and African Helmeted Turtle. Madagascar's rich biodiversity and high endemism across various taxa make it critical for conservation and scientific research.
3. **The Democratic Republic of Congo** contains the second-largest jungle in the world after the Amazon rainforest, and it is the largest on the African continent. This country boasts a rich variety of flora and fauna, comprising various species, including 900 different species of butterflies, 400 mammalian species, 280 reptile species, and many other endemic species such as the Bonobo and those found in the Congo River Basin.

Oceania

1. **Australia**, an isolated continent-nation, hosts a wide range of animal species, from the iconic kangaroos and koalas to smaller, more dangerous creatures. Kangaroos, found only in Australia, are part of the marsupial family. As one of just two developed nations with mega-biodiversity, Australia is home to an estimated 600,000 to 700,000 plant and animal species. This includes over 378 mammal species, 828 bird species, 300 lizard species, and 140 snake species. The country ranks first globally in endemic species, such as kangaroos, dingoes, and koalas.
2. **Papua New Guinea**, located within the Pacific Ring of Fire, is characterized by its numerous active volcanoes, coral formations, dense rainforests, and diverse marine and terrestrial ecosystems. This country,

less industrialized than its neighboring Indonesia, boasts vast areas of untouched wilderness. It is famous for its unique flora and fauna, including the world's third-largest rainforest. The nation's indigenous wildlife features unusual species like carnivorous mice, tree-climbing kangaroos, and giant rodents. Furthermore, the island is inhabited by 730 bird species, more than 320 amphibian species, and roughly 300 different reptiles.

Biodiversity hotspots

Norman Myers introduced the concept of "biodiversity hotspots." These are areas with high species diversity that face significant extinction threats due to overexploitation of natural resources or other human activities, such as wasteful use of natural resources, deforestation, poaching, and hunting. This biodiversity is crucial for all living organisms' survival, as it regulates climate conditions, provides essential products necessary for life on Earth, and helps meet our basic needs. There are 36 biodiversity hotspots worldwide, with 4 located in India: the Himalayas, Indo-Burma Region, Western Ghats, and Sundarbans. The remaining 32 hotspots include the Afghanistan Mountains, Alaska Boreal Forests, Atlantic Forest, Cape Floristic Region, Caribbean Islands, Chilean Winter Rainfall-Valdivian Forests, Congo Basin, East Melanesian Islands, Eastern Afromontane, Guinean Forests of West Africa, Himalaya, Indo-Burma, Indonesia, Japan, Madagascar and the Indian Ocean Islands, Mediterranean Basin, Mesoamerica, Mountains of Central Asia, Mountains of Southwest China, New Caledonia, New Zealand, North American Coastal Plain, North American Deserts, the Philippines, Polynesia-Micronesia, Sahara and Sahel, Southwest Australia, Sundaland, Tehuacán-Cuicatlán Valley, Tropical Andes, Tropical Asia, Tumbes-Chocó-Magdalena, U.S. California Floristic Province, Venezuelan Coastal Cordillera, Western Ghats and Sri Lanka, and Yunnan-Sichuan. Preserving these biodiversity hotspots is essential for maintaining global biodiversity. It also safeguards genetic diversity, which aids in adaptation to environmental changes.

IUCN Threat Categories

The International Union for Conservation of Nature (IUCN), established in 1948, is a worldwide organization that unites various NGOs, governmental bodies, and individuals to protect nature and promote sustainable practices for preserving Earth's biodiversity. Renowned for its Red List of Threatened Species, the IUCN evaluates the extinction risk of numerous species globally, guiding conservation efforts and policies worldwide. The IUCN has defined seven distinct threat categories:

1. **Least Concern Species:** Organisms with minimal risk of extinction due to their abundant population. Examples include American black bears, sharks, and sawfish.
2. **Vulnerable Species:** Creatures likely to become endangered if harmful factors persist. Box turtles and tigers fall into this category.
3. **Endangered Species:** Animals on the brink of extinction, such as snow leopards, Asiatic lions, and whooping cranes.
4. **Extinct Species:** Organisms that no longer exist, like the Spix's macaw and golden toad.
5. **Critically Endangered:** Species facing an extremely high risk of eradication, including the Northern white rhinoceros and brown spider monkey.
6. **Near Threatened:** Animals close to being classified as threatened, such as blue sharks and ball pythons.
7. **Data Deficient:** Species lacking sufficient information for a proper assessment of their extinction risk, like orcas and hand fish.
8. **Not Evaluated:** Organisms that have not yet been assessed to determine their risk of disappearance, such as Eurycea aquatica and milkfish.

Red Data Book

The IUCN (International Union for Conservation of Nature) Red List, also referred to as the RED DATA BOOK, is a publication by IUCN aimed at raising awareness and encouraging the preservation of biodiversity. The book offers in-depth assessments of numerous species, encompassing their population dynamics, geographical range, habitat needs, and potential threats. Species are categorized from 'Least Concern' to 'Extinct' based on their extinction risk. This resource is regularly updated with current information on species status worldwide. As of the 2024 update, the IUCN Red List has evaluated 157,000 species, with 45,300 considered threatened, falling into the categories of Critically Endangered (CR), Endangered (EN), and Vulnerable (VU). While this represents significant progress, continued efforts are necessary to safeguard these species and their habitats. The Red data book also emphasizes ongoing challenges faced by various species due to factors such as habitat destruction, climate change, and human impact. Amphibians are among the most at-risk groups, with 41% of assessed species threatened. Climate change and ocean acidification have led to 36% of reef corals being threatened. The Bornean Elephant is now classified as Endangered due to habitat loss and human activities.

Importance of Biodiversity

1. **Financial Advantages:** Ecosystems rich in biodiversity contribute to economic growth through sectors like farming, fisheries, and travel. They provide income sources for countless individuals.
2. **Heritage Significance:** Numerous societies have profound connections with nature and depend on biodiversity for customary practices and spiritual fulfillment.
3. **Climate Moderation:** Varied ecosystems assist in regulating climate by absorbing carbon dioxide and generating oxygen. Woodlands, marshes, and oceans play a vital role in alleviating climate change.
4. **Economic Worth:** Many sectors, including agriculture, pharmaceuticals, and tourism, rely on biodiversity. Its conservation ensures the longevity of these economic activities.
5. **Cultural Importance:** Many societies have deep-rooted connections to biodiversity, relying on it for traditions, livelihoods, and spiritual well-being.
6. **Inherent Value:** Each species possesses an intrinsic worth and the right to exist. Preserving biodiversity is also an ethical duty to safeguard the planet's natural legacy.
7. **Water Purification:** Wetlands and other ecosystems filter contaminants from water, ensuring clean and safe water for human and animal consumption.
8. **Disaster Protection:** Biodiverse coastal ecosystems like mangroves and coral reefs shield coastlines from storms and erosion, mitigating the impact of natural disasters.
9. **Genetic Resources:** Biodiversity offers a pool of genetic material that can be crucial for developing new crops and livestock resistant to diseases and climate change.
10. **Aesthetic and Recreational Worth:** Natural landscapes and wildlife provide recreational, tourism, and aesthetic experiences, contributing to human well-being and economic activity.

Threats to biodiversity

Human activities pose numerous threats to biodiversity, with population growth being a primary factor. As the number of people increases, so does the demand for essential resources like food, housing, and jobs. To satisfy these needs, humans often damage ecosystems, leading to the destruction of forests and the transformation of wooded areas into farmland. The expanding population›s need for living space results in further deforestation. Many species face extinction due to the unsustainable exploitation of natural resources. Various human actions, including hunting, clearing land, poaching, and overfishing, contribute to declining species populations. Urban development fragments habitats, making it difficult for species to find mates. The conversion of forests into agricultural or urban areas causes numerous species to lose their homes. Industrial and vehicular pollution harms various plant species, while the repurposing of forested land displaces many animal species. These diverse human activities collectively endanger biodiversity.

Ecosystem Services

The ecosystem offers numerous advantages that sustain human life. The diverse array of services provided by ecosystems is crucial for the survival of living organisms. Here's a brief overview of these services:

1. Cultural services

1. Ecosystems deliver various intangible benefits, including tourism, leisure activities, and relaxation.
2. Many societies derive their cultural identities and customs from their natural environment, often maintaining spiritual and religious ties to nature.
3. The natural world provides opportunities for education and scientific investigation.

2. Provisioning services

1. Nature supplies timber and fiber, which are utilized for various purposes such as construction, clothing, and paper products.
2. Wild species contribute to agricultural and medical advancements by offering genetic material for breeding and research.
3. A thriving ecosystem yields a variety of food products, including fruits, fish, livestock, grains, vegetables, and medicinal plants.

3. Regulation services

1. Ecosystems play a vital role in air quality regulation, as natural elements absorb pollutants. Forests and oceans absorb CO2, aiding in climate regulation.
2. Forests take in carbon dioxide and release oxygen, which is essential for respiration. They also remove pollutants from the air, enhancing air quality.

3. Wetlands and forests filter contaminants from water and act as sponges, absorbing excess rainwater, which helps regulate flooding.

4. Supporting services

1. Soil health is maintained through nutrient cycling, which enhances soil fertility, provides land for agriculture or vegetation, and combats desertification.
2. Plants transform sunlight into energy through photosynthesis, forming the foundation of the food web.
3. Diverse ecosystems ensure the stability of the planet's biological networks, supporting life at all levels. Wetlands and forests are crucial in maintaining the water cycle, ensuring fresh water for drinking, crop cultivation, and wildlife sustenance.

Ways to conserve biodiversity

By now, we understand the necessities of conserving the ecosystem. Now let's discuss the steps taken by IUCN (International Union for Conservation of Nature) to conserve biodiversity.

IUCN identifies two ways to conserve biodiversity:

1. Ex-Situ Conservation
2. In-Situ Conservation

1. **In-Situ Conservation** - The practice of safeguarding an entire ecosystem's biodiversity is termed In-Situ Conservation, which refers to on-site preservation. This approach protects all species within their natural environments, facilitating sustainable land management and ecosystem restoration. It enables species to persist in their native habitats and maintains genetic diversity across various organisms. In-situ conservation is particularly vital for preserving endemic species and is a crucial step in protecting biodiversity-rich areas. Globally, there are over 6,555 national parks and 759 biosphere reserves spanning 7,442,000 sq. km across 136 countries. Additionally, more than 16,615 marine protected areas exist worldwide, though these only cover 7.2% of the world's oceans. India specifically hosts 573 wildlife sanctuaries, 106 national parks, 31 marine protected areas, and 18 biosphere reserves.
2. **Ex-Situ Conservation** - When a particular plant or animal faces extinction risk or is classified as a threatened species, it undergoes ex-situ conservation, which involves relocation to a specialized environment outside its natural habitat. In these carefully managed

settings, the organism receives dedicated care to prevent its eradication. Additionally, the gametes of these organisms are biologically preserved to maintain their populations in future generations. Animal gametes are cryopreserved and conserved through various genetic engineering techniques, while plant species are maintained using advanced tissue culture methods. This approach not only prevents species extinction but also ensures the continuation of their populations. Examples of ex-situ conservation include seed banks, zoos, cryobanks, wildlife safari parks, and botanical gardens.

Protected Areas and Functions

Protected areas are essential for conserving biodiversity rich regions and maintenance of ecosystem. Here we discuss about several conservation areas, selected for conserving planets biodiversity.

1. **Mawphlang Sacred Grove, situated in Meghalaya, India:** This area harbors numerous rare and endangered plant species. The Khasi tribe has protected these lands, strictly limiting hunting and logging activities to ensure the preservation of species.
2. **Kavu, another sacred grove found in Kerala, India:** This region is deeply associated with deities and serves as a vital ecological niche for conserving indigenous flora and fauna.
3. **Great Barrier Reef Marine Park, located in Australia:** This marine sanctuary is home to thousands of aquatic species. Fishing and tourism activities are heavily restricted in this area.
4. **Congo Basin Rainforest, situated in central Africa:** It is known for housing gorillas, elephants, and various plant species, this region prohibits ecosystem-damaging activities such as poaching, deforestation, and illegal logging.
5. **Galápagos National Park, Ecaudor:** This park is famous for Darwin's theory of evolution. This park strictly regulates tourism, fishing, and scientific research activities.
6. **Serengeti National Park, Tanzania:** It is renowned for its annual wildebeest migration. This park supports diverse wildlife, including lions, elephants, and cheetahs. Conservation efforts include anti-poaching measures, habitat restoration, and community-based programs.

Very Short Answer Type Questions

1. Explain the term biodiversity?
2. Illustrate different types of biodiversity.

3. What do you understand from the term species richness and species evenness?
4. Who is the founder of the term biodiversity?
5. Who gave the concepts of biodiversity hotspots?
6. Explain the term sacred groves.
7. What is meant by endemic species?
8. Define Ex- situ, with examples.
9. Define In- Situ, with examples.
10. What is the importance of biodiversity?

Short Answer Type Questions

1. What are biodiversity hotspots? Who coined this term and why?
2. What are the biodiversity conservation strategies employed by IUCN?
3. What is in Red Data Book?
4. What are IUCN threat categories?
5. Explain species richness and evenness along with their examples.
6. Give examples of threatened species.
7. Differentiate between threatened and endangered species.
8. Discuss the importance of biodiversity.
9. Explain the key features of Red data book.
10. Tell the significance of biodiversity conservation.

Long Answer type questions

1. Discuss mega diversity zones in detail, also discuss all megadiverse countries.
2. Which continent exhibits biodiversity richness, discuss about that continent in detail also mention all the biodiverse countries it holds in it.
3. Discuss the various biodiversity hotspots.
4. What is IUCN? Discuss all eight threat categories of IUCN.
5. What is the significance of biodiversity. What are the essential steps can be taken to conserve biodiversity.
6. Discuss the significance of biodiverse ecosystem.
7. Discuss different biodiversity conservation ways and name some biodiversity protected areas.

Multiple Choice Questions

1. How many total megadiverse countries are there in world-wide level?
   a. 12 countries
   b. 9 countries
   c. 7 countries
   d. 17 countries
2. Box turtles and handfish are examples of
   a. Data deficient and endangered species.
   b. Data deficient and Vulnerable species.
   c. Vulnerable species and endangered species.
   d. None of them.
3. What are the threats of biodiversity?
   a. Deforestation
   b. Resource depletion
   c. Poaching
   d. All of the above.
4. Congo Basin Rainforest is an example of
   a. Megadiverse zone
   b. Sacred grove.
   c. National Park
   d. Marine Park.
5. How many biodiversity hotspots exist in India.
   a. Four
   b. Six
   c. Only two
   d. None of the above
6. Deep rooted connection with nature represent which ecosystem services.
   a. Provisioning services.
   b. Regulation services
   c. Supporting services
   d. Cultural services.
7. Preservation of gametes in cryobanks represents which type of conservation
   a. *Ex- situ*
   b. *In- situ*
   d. Both of them
   d. Cryobanks
8. Total no. of national parks observed in India.
   a. 5000
   b. 136
   c. 759
   d. 106
9. Who coined the term biodiversity hotspots
   a. E.O Wilson
   b. Charles Darwin
   c. Norman Myers
   d. None of the above.
10. Which nation is renowned as Big Five Wildlife-
    a. India
    b. US
    c. Africa
    d. Australia

5

Natural Resources

Water Resources

Water Availability

From an early age, we're taught that 70% of Earth's surface is water and 30% is land, yet numerous regions worldwide struggle with water scarcity. The reason for this paradox lies in the composition of Earth's water: 96.0% is saline ocean water, nearly 1% is found in saline lakes and groundwater, and only around 3% exists as freshwater. This limited freshwater supply is often used unsustainably by humans, exacerbating its scarcity. Activities such as laundering clothes in rivers, leaving faucets running, and misusing water contribute to the shortage of clean, fresh water. Interestingly, even places like Cherapunji and Mawsynram, known for receiving the world's highest rainfall, experience water shortages during dry seasons.

India's river system, another source of drinking water, includes many rivers considered sacred. These waterways are categorized as either perennial or seasonal. Melting glaciers play a crucial role in providing freshwater resources, particularly during lean periods. The water from melting snow yields approximately twice the amount compared to South Peninsular rivers. Rivers such as Godavari, Narmada, Krishna, and Cauvery receive water primarily during rainy seasons. Groundwater aquifers store water during rainy periods and release it during dry spells.

Clean, fresh water is essential for various purposes, including household use, irrigation, and industrial applications. Despite having abundant freshwater resources, many regions still face challenges in accessing fresh water. This issue stems from unsustainable water usage, poor water management practices, and environmental degradation.

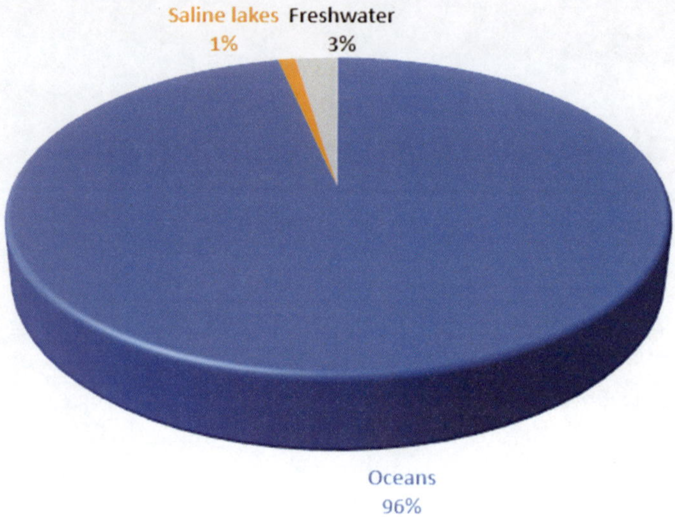

Figure 7: Availability of Water on Earth

Quality Aspects

Water is crucial for life on Earth, with humans unable to survive beyond 3-4 days without it. A 15% loss of water can be fatal for organisms. The unavailability of clean drinking water, inadequate sanitization, and poor hygiene practices lead to various diarrheal diseases, claiming numerous lives, particularly among children under 5. Many water sources become polluted due to harmful chemicals from agricultural runoff, industrial waste, and mining activities, which contribute to the deterioration of global water quality. The World Health Organization reported that in 2008, about 880 million people worldwide (13% of the global population) lacked access to safe drinking water. The water crisis is exacerbated by social inequalities, with impoverished individuals more likely to lack clean water and sanitation compared to their wealthier counterparts in similar regions. On a global scale, enhancing water safety, sanitation, and hygiene could prevent up to 9% of all diseases and 6% of all deaths. In the United States, over 40,000 water bodies currently meet the EPA's (Environmental Protection Agency) definition of "impaired," indicating their inability to support healthy ecosystems or meet water quality standards. Water pollutant concentrations are typically measured in minute units, such as parts per million (ppm) or parts per billion (ppb). For instance, 1 ppm of arsenic equates to one part arsenic per million parts water, or one drop of arsenic in 50 litres of water. To better comprehend these small concentration units, 1 ppm can be visualized as 1 cm (0.4 in) in 10 km (6 miles) in terms of length, or 30 seconds in a year in terms of time.

Waterborne Disease

Waterborne pathogens are responsible for the transmission of various illnesses. These ailments primarily occur when individuals consume contaminated food or water, or come into direct contact with tainted water sources. Such exposure can result in numerous waterborne diseases, some of which may pose life-threatening risks. This significant health issue necessitates the implementation of crucial preventive measures. Several examples of waterborne diseases are outlined below:

1. **Cholera:** This illness is caused by *Vibrio cholerae* bacteria. It spreads through consuming water contaminated with feces containing the bacteria. The symptoms include vomiting, diarrhea, and potential complications like dehydration.
2. **Typhoid:** *Salmonella typhi* bacteria cause this disease. Infection occurs when a person ingests food or water contaminated with feces from an infected individual. Typical symptoms include fever, headache, stomach pain, loss of appetite, and fatigue.
3. **Giardiasis:** This disease is spread through water contaminated with parasites, specifically *Giardia lamblia*. Affected individuals may experience abdominal cramps, nausea, flatulence, and diarrhea.
4. **Dysentery:** *Shigella* bacteria or *Entamoeba histolytica* parasites are responsible for causing dysentery. Transmission occurs through contaminated water and food sources. Symptoms typically include bloody diarrhea, fever, and abdominal cramps.
5. **Hepatitis A:** The *Hepatitis* A virus is the causative agent of this disease. Infected individuals may experience weakness, fever, jaundice, abdominal pain, and nausea.
6. **Schistosomiasis:** This condition is caused by *Schistosomiasis* parasites. Transmission occurs through contact with water containing infected freshwater snails. The disease can lead to abdominal pain, blood in urine, diarrhea, and liver enlargement.

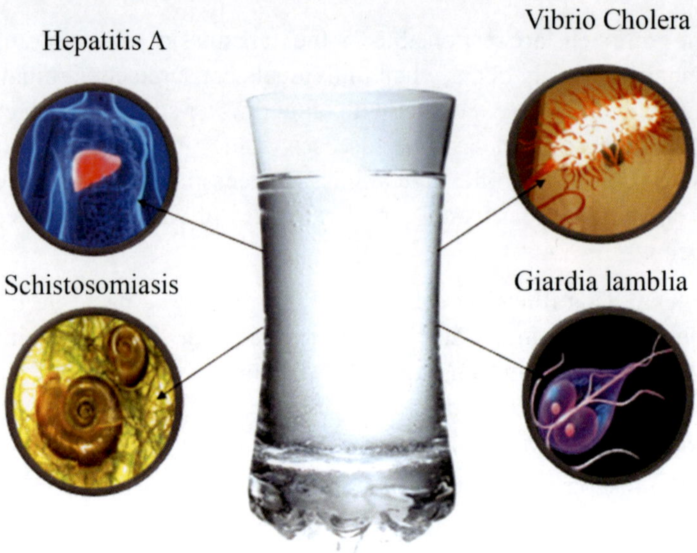

Figure 8: Waterborne Pathogens

Preventive Measures of Water Borne Disease

Waterborne illnesses pose a significant risk to everyone's health and can be fatal. Therefore, it's crucial to implement preventive measures against these dangerous diseases, which come with numerous health risks. Here are some strategies to safeguard against waterborne illnesses:

1. Steer clear of water sources known to be contaminated.
2. Educate people about the dangers of waterborne diseases.
3. Implement proper sanitation methods.
4. Get appropriate vaccinations to boost protection against these illnesses.
5. Purify drinking water by boiling or filtering it before consumption.
6. Refrain from consuming food sold on the street or in open-air markets.
7. Maintain good personal hygiene.

Water-Induced Diseases

Water induced diseases are Diseases related to water are ailments that result from water-associated factors, rather than directly from pathogens in the water. These illnesses are frequently associated with environmental conditions and organisms that flourish in or near aquatic environments.

1. **Malaria:** The *Plasmodium* parasite, which causes malaria, is transmitted by Anopheles mosquitoes that breed in stagnant water. The offspring of these mosquitoes spread the disease, resulting in symptoms such as headache, fever, chills, fatigue, and muscle pain.
2. **Dengue fever:** *Aedes* mosquitoes, which transmit dengue, proliferate in water-holding containers like flower pots and coolers, especially in areas with poor sanitation. Symptoms of dengue include high fever, eye pain, joint and muscle aches, and severe headache.
3. **Leptospirosis:** This disease is caused by *Leptospira* bacteria found in contaminated animal urine. Symptoms include jaundice, muscle aches, chills, vomiting, and fever.
4. **Lymphatic Filariasis (Elephantiasis):** The parasites responsible for this condition spread through bacteria present in stagnant water. The disease manifests with symptoms such as limb swelling, skin thickening, and genital swelling.

Precautions

1. Eliminate standing water sources.
2. Practice proper water management and maintain cleanliness in your surroundings.
3. Do protective clothing that covers your arms or utilize mosquito netting.
4. Promote public awareness through educational initiatives and online advertisements.

Fluoride Problem in Drinking Water

The presence of fluoride in drinking water offers both advantages and potential drawbacks, contingent on its concentration and exposure duration. Fluoride strengthens dental enamel, enhancing its resistance to decay. It can also counteract early tooth decay by restoring weakened enamel. Research indicates that fluoridated drinking water decreases cavity occurrence in both young and adult populations. Fluoridation of water supplies is a cost-efficient method to enhance dental health, particularly in areas with limited access to dental services. The U.S. Public Health Service suggests an optimal fluoride level of 0.7 milligrams per litre of water for dental health benefits. Levels exceeding 4.0 ppm (parts per million) are deemed unsafe. Overexposure to fluoride during tooth development in childhood can result in dental fluorosis, characterized by white marks or spots on the enamel. While usually mild and aesthetic, this condition can be more severe in some instances. Prolonged exposure to high fluoride concentrations may lead to skeletal fluorosis, causing pain and damage to bones and joints. Elevated fluoride levels can impact thyroid function, potentially triggering hyperparathyroidism, which involves unregulated secretion of parathyroid hormones and can result in abnormally high blood calcium levels.

While fluoride in drinking water is generally considered safe and beneficial at recommended concentrations, excessive exposure can pose health risks. It is crucial to monitor fluoride levels and adhere to guidelines to ensure safe consumption.

Mineral Resources

Minerals are naturally occurring inorganic substances found in the Earth's crust. They serve various purposes, including industrial applications, jewellery creation, and construction. Additionally, minerals play a crucial role in technological progress, economic growth, infrastructure development, and agriculture. There are several categories of mineral resources:

1. **Metallic minerals:** These are minerals from which metals can be extracted, such as gold, silver, iron, copper, and aluminium. They are utilized for numerous purposes, primarily in jewellery design, electrical wiring, packaging, construction, and transportation.
2. **Non-metallic minerals:** This category includes minerals that do not yield metals. They are mainly used in the production of construction materials like bricks, cement, plaster, and ceramics. Examples include clay, gypsum, and limestone.
3. **Energy minerals:** These minerals produce energy and include coal, petroleum, and uranium. Coal and natural gas are utilized for electricity generation, while petroleum and uranium are used to produce fuel.

Various methods are employed to extract minerals, depending on the specific mineral type. Some common extraction techniques include:

1. **Quarrying:** This process begins with selecting a suitable site that has the potential for maximum yield. Explosives are used to break rocks into smaller pieces, which are then crushed, screened, and washed. Minerals extracted through quarrying include sand, gravel, stone, limestone, and slate.
2. **Underground mining:** This method is used to extract minerals located deep beneath the Earth's surface. It involves creating underground tunnels or shafts to access the minerals. E.g.- Lead, silver, platinum, copper.

3. **Open-pit mining:** Also known as open-cast or open-cut mining, this surface mining technique is used to extract minerals and ores near the surface. It is employed when the ore body is close to the surface and covers a large area, making it more economical and safer than underground mining. E.g.- Bauxite, phosphate, limestone
4. **Placer mining:** This method is used to extract valuable minerals from placer deposits, which are mineral concentrations found in sediments of riverbeds, stream beds, and other alluvial environments. It is particularly effective for extracting gold, gemstones, and other heavy minerals that have eroded from their primary sources and been transported by water.
5. **Leaching:** This process involves using chemicals to dissolve valuable minerals from ores, which can then be recovered from the resulting solution. E.g.- Uranium, Lithium, Zinc, Nickel.

Forest Wealth

Forest wealth encompasses the economic and social advantages derived from forest ecosystems. The resources extracted from forests enhance our daily lives. Forests are home to a diverse array of plant and animal species, providing essential products for the survival of Earth's inhabitants. Humans rely on forests not only for raw materials but also for crucial elements of environmental health and sustainability. The benefits of forest health can be categorized as follows:

1. **Social wealth:** Forests contribute significantly to human mental well-being by offering clean air, spaces for relaxation and meditation, and a connection to nature. This natural environment promotes peace and tranquillity, helping individuals cope with various health issues such as depression, anxiety, stress, and loneliness. Additionally, forests provide opportunities for scientists to conduct environmental education research, which can have both positive and negative impacts on the environment.
2. **Economic wealth:** Forests yield numerous timber and non-timber products utilized in construction, manufacturing, industrial processes, and the development of various medicines. They also generate employment opportunities through the harvesting and processing of these products. Furthermore, forests attract tourists and serve as recreational destinations.
3. **Cultural wealth:** Many communities depend entirely on forests for their livelihoods. Certain endemic plant species found in forests play a crucial role in maintaining biodiversity.

4. **Ecological health:** Forests aid in carbon sequestration, as a substantial number of plants absorb carbon dioxide and release oxygen, benefiting human health. They also play a vital role in water regulation, contributing to timely rainfall and maintaining water quality and availability. Moreover, forests help enhance soil fertility.

In conclusion, forest wealth significantly impacts the lives of organisms. Therefore, it is essential to utilize these resources sustainably to ensure their long-term preservation and benefits.

Material Cycle

Material cycles describe the movement of substances such as carbon, nitrogen, sulfur, phosphorus, and water within Earth's ecosystem. These processes involve the creation, transformation, and reintegration of materials in biological systems, ensuring their sustainability. These cycles are crucial for life-sustaining processes like respiration, photosynthesis, and decomposition. Several key material cycles are outlined below:

1. **Carbon cycle**: Industrial waste and vehicle emissions release significant amounts of carbon into the atmosphere. Plants absorb this atmospheric carbon dioxide during photosynthesis. Herbivores then consume these plants. When organisms die, their remains and waste products decompose in the soil. Over millions of years, this organic matter transforms into fossil fuels, which are later used in industrial processes and to create petroleum products. This process maintains the continuity of the carbon cycle.

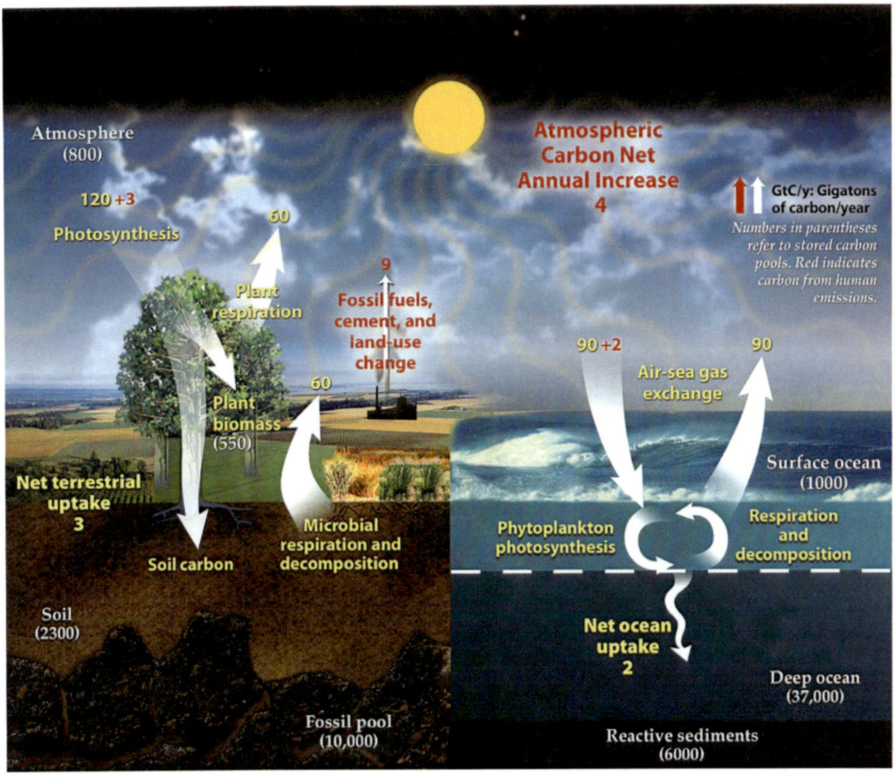

Figure 9: Carbon Cycle

2. **Nitrogen cycle:** This biogeochemical cycle involves the conversion of nitrogen into various chemical forms. The atmosphere contains abundant nitrogen, which plants absorb. Bacteria in plant roots convert this atmospheric nitrogen into ammonia, making it usable for plants - a process called nitrogen fixation. Decomposing organic matter and animal waste provide organic nitrogen, which bacteria like *Pseudomonas* and *Bacillus* convert into ammonia through ammonification. While some plants can use nitrogen in its ammonium form, others cannot due to its toxicity. For these plants, ammonium is converted into nitrite and nitrate through nitrification, a process facilitated by bacteria such as *Nitrosomonas* and *Nitrococcus*. Subsequently, nitrites and nitrates are transformed back into nitrogen through denitrification, releasing nitrogen into the atmosphere and completing the cycle.

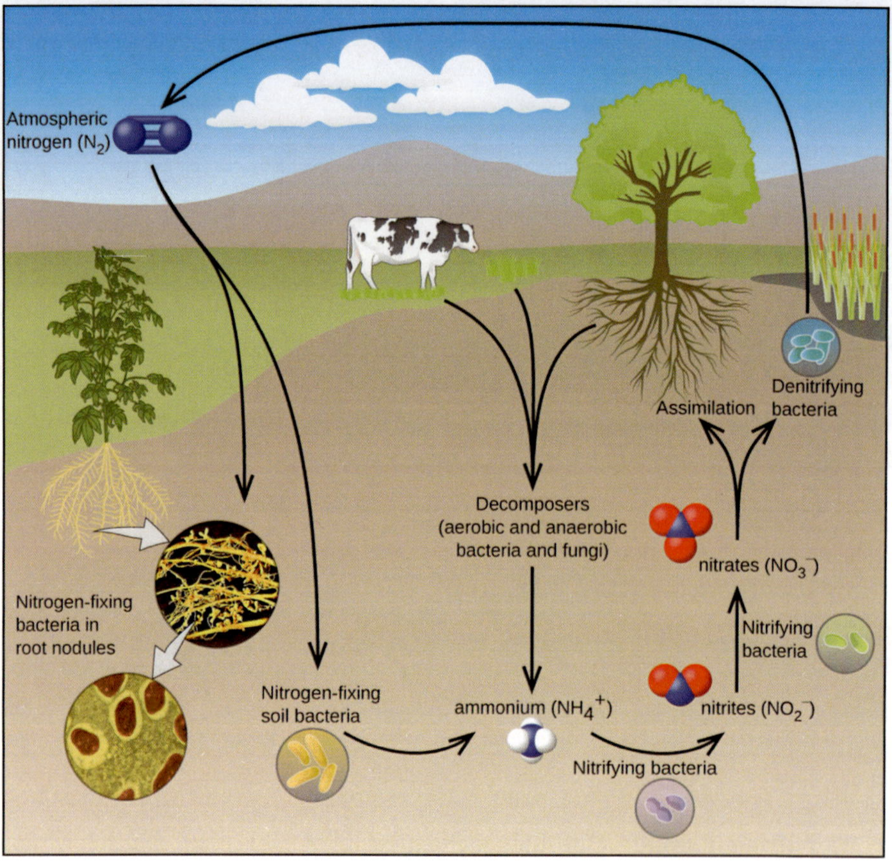

Figure 10: Nitrogen cycle

3. **Sulfur cycle:** This cycle illustrates the ongoing movement of sulfur between rocks, water, plants, and animals in the environment. Sulfur exists in various forms, including sulfate (SO_4) and hydrogen sulfide (H_2S). Volcanic eruptions and organic matter decomposition release sulfur into the atmosphere, which then enters soil and water systems as acid rain. Microorganisms play a vital role in converting sulfur into different chemical forms, making it available for biological use and returning it to the atmosphere.

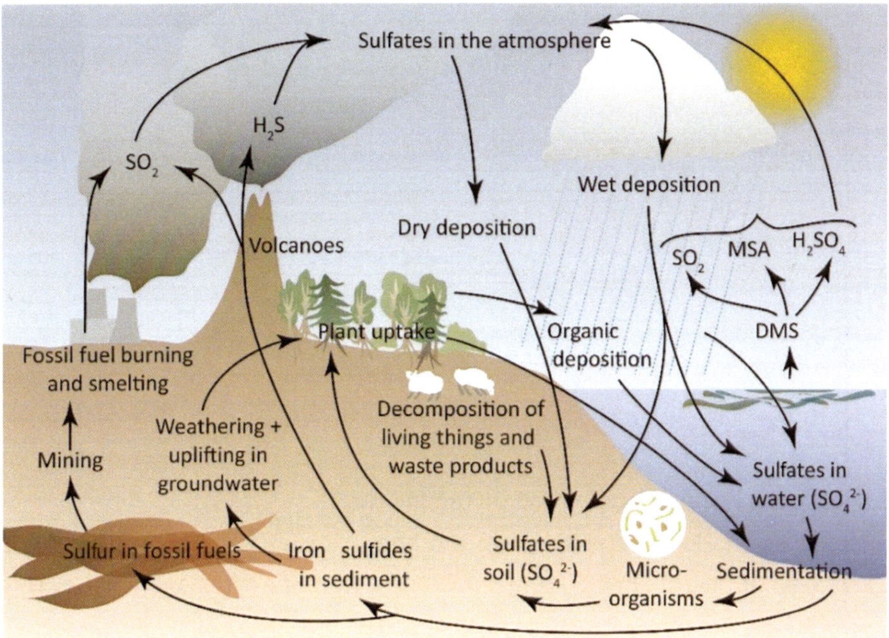

Figure 11: Sulphur Cycle

1. Very Short Answer Type Questions

1. Differentiate between perennial and seasonal rivers.
2. Define water borne diseases and water induced diseases.
3. Mentions the precautions should be taken against water induced diseases.
4. Mention the side effects of taking fluoride in drinking water.
5. Discuss about different types of mineral resources.
7. Define them: Quarrying, open pit mining, underground mining.
6. Discuss social wealth.
7. What are material cycles?
8. What is forest wealth?
9. Give some examples of minerals extracted through open pit mining activities

Short Answer Type Questions

6. Differentiate between water borne diseases and water induced diseases.
7. Discuss *Schistosomiasis* and *Giardiasis* disease.
8. Suggest some ways to prevent water borne diseases.

9. Give examples of any two water induced diseases.
10. Discus the problems caused due to presence of high concentration of fluoride in water.
11. Define mineral extraction methods.
12. Discuss nitrogen cycle.
13. Discuss material cycle with examples.

Long Answer Type Questions

4. Discuss the problem faced in availability of clean drinking water.
5. Why water borne diseases are the serious health concern?
6. What precautions one can take to be protected from water induced diseases.
7. Discuss mineral extraction method in detail.
8. Describe the benefits of forest wealth.
9. Give a diagrammatic representation of Sulphur cycle and discuss it in detail.
10. Discuss several water induced diseases.

Multiple Choice Questions

1. Krishna and Godavari are-
   - **a. Perennial rivers**
   - b. Seasonal rivers
   - c. Perennial and seasonal rivers
   - c. None of the above.
2. Total amount of water covered by ocean-
   - a. 97%
   - **b. 96.5%**
   - c. 97.5%
   - d. 3%
3. Cholera, typhoid and malaria are-
   - a. Water borne diseases.
   - b. Water induced diseases.
   - d. Water borne and water induced diseases
   - **c. None of them.**
4. Bauxite, phosphate, limestone can be extracted through which mining method-
   - a. Open cut mining
   - b. Underground mining
   - c Open cast mining
   - **d. Both a & c**
5. Clay, gypsum, limestone is an example of –
   - a. Minerals.
   - b. Mettalic minerals
   - **c. Non- mettalic minerals**
   - d. None of the above.

6. Tourism and recreation is a part of –
   a. Economic wealth
   b. Social wealth
   c. Ecological wealth
   d. Cultural wealth
7. Ammonification means-
   a. Conversion of ammonium into nitrites.
   b. conversion of atmospheric nitrogen into ammonia
   c. transformation of organic nitrogen into ammonia
   d. both b & c.
8. Nitrosomonas & pseudomonas are the examples of –
   a. Nitrogen fixation
   b. Denitrification
   c. Ammonification
   d. Ammonification and nitrification.
9. Which mineral is used in designing bricks, cement, ceramics
   a. Non- metallic minerals
   b. Clay
   c. Gypsum
   d. Clay and limestone.
10. Which type of minerals are used in designing jewwelery and electrical wirings
    a. Non mettalic minerals
    b. Energy minerals
    c. Mettalic minerals
    d. Both b & c.

Section III
Ecosystem Processes and Energy Dynamics

6

Energy and Environmental Impact

The Industrial Revolution marked a pivotal shift in energy production, transitioning from manual labour, animal power, and natural sources to the widespread use of fossil fuel combustion. This transformation, accounting for over 86% of global energy consumption, has resulted in significant enhancements in living standards across industrialized nations. A substantial portion of the world's population now enjoys access to abundant energy at seemingly affordable prices. However, this progress has come at a considerable environmental cost, stemming from the dwindling supply of fossil fuels and the inefficient conversion of energy into less useful forms.

Energy, a multifaceted and crucial subject, permeates various aspects of our daily lives and the world at large. Defined as the capacity to perform work, energy manifests in diverse forms and is primarily categorized into two types: primary and secondary. Primary energy is directly extracted from the environment, while secondary energy results from the conversion of primary sources into electricity or fuel. Primary energy is further classified into renewable and non-renewable sources.

Different Types of Energy

Renewable energy, often termed **non-conventional**, encompasses naturally replenishable and less polluting sources. These infinite resources, such as solar, hydrothermal, geothermal energy and others are becoming increasingly efficient and cost-effective due to technological advancements. Conversely, **non-renewable or conventional energy** sources, including nuclear energy and fossil fuels, are finite and will eventually be exhausted. While these sources provide substantial energy per unit and offer consistent, controllable power, their limited availability necessitates sustainable usage to meet future needs. Energy also exists in other forms, such as kinetic energy (associated with particle motion) and potential energy (related to particle positioning). Additional examples include thermal, magnetic, tidal, and nuclear energy.

Electromagnetic Radiation

Electromagnetic radiation (EMR) is energy released by charged particles, manifesting as waves or particles moving through space. It is characterized by electric and magnetic fields oscillating perpendicularly to each other and to the wave propagation direction. EMR covers a vast range of frequencies and wavelengths, from low-energy radio waves to high-energy gamma rays. The electromagnetic spectrum is segmented into categories based on frequency, including radio waves, microwaves, infrared radiation, visible light, ultraviolet radiation, X-rays, and gamma rays. Each category has distinct properties and applications; for example, visible light is crucial for vision, microwaves are used in communication and heating, and X-rays are indispensable for medical imaging. Although EMR is essential to many natural and technological processes, excessive exposure to high-energy radiation can lead to health risks, including tissue damage and DNA mutations. The study and application of EMR have facilitated advancements in communication, healthcare, and energy, establishing it as a fundamental element of contemporary science and technology. Electromagnetic radiation, traveling through space at the speed of light, comprises oscillating electric and magnetic fields. This radiation spans a broad spectrum of wavelengths and frequencies, which are inversely related. The energy of electromagnetic radiation is directly proportional to its frequency, with higher frequency radiation possessing more energy. The electromagnetic spectrum is organized based on wavelengths and frequencies. Cosmic rays exhibit the highest energy and shortest wavelength, whereas radio waves have the longest wavelength and lowest energy. Seven distinct types of electromagnetic radiation are represented in the accompanying diagram:

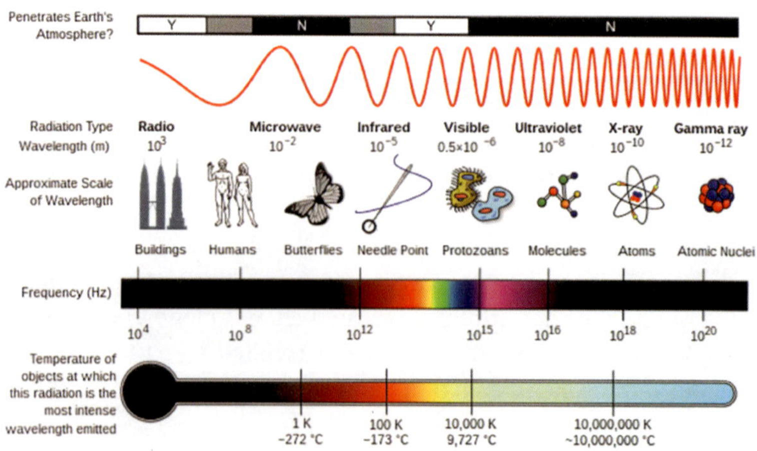

Figure 12: Electromagnetic Radiations

Applications of Electromagnetic Radiations

1. Radio waves and microwaves play a pivotal role in broadcasting, satellite communication, and wireless networks.
2. X-rays and gamma rays offer non-invasive techniques for internal body examination.
3. Ultraviolet (UV) radiation is employed in disinfection processes, while gamma rays are utilized for sterilizing medical equipment and food items.
4. High-energy electromagnetic radiation, including UV light, X-rays, and gamma rays, has the potential to ionize atoms, potentially leading to biological tissue damage and elevated cancer risk.
5. Ongoing advancements in electromagnetic technology continue to transform fields such as communication, medicine, and astronomy.

Conventional sources or Non-Renewable sources

These finite resources have been instrumental in industrial development for centuries. Conventional sources encompass fossil fuels, including coal, oil, and natural gas.

Fossil Fuel-Based Sources

Fossil fuels are energy sources derived from the decomposed remains of ancient flora and fauna, which have been buried and subjected to intense heat and pressure over geological timescales. The three primary categories of fossil fuels are natural gas, coal, and oil.

1. Coal

Coal is a combustible sedimentary rock, black or brownish-black in colour, formed from the remains of ancient plant material. Its primary constituent is carbon, accompanied by various elements such as hydrogen, sulfur, oxygen, and nitrogen. Coal serves as a fuel source for electricity generation, industrial processes, and heating applications. The formation of coal occurs over millions of years through the accumulation and burial of plant material in swampy environments. This process begins with the accumulation of plant material in wet, anaerobic conditions, forming peat, the precursor to coal. As pressure and heat increase over time, peat transforms into lignite, a low-grade, brownish coal with high moisture content. Further compaction and heating convert lignite into sub-bituminous coal, characterized by lower moisture content and higher carbon content. Through continued exposure to heat and pressure, sub-bituminous coal undergoes transformation into bituminous coal, a prevalent energy source characterized by its higher energy density. The culmination of coal formation results in anthracite, a hard, black coal with increased carbon content and energy density, which combusts more cleanly than other coal varieties. Coal classification is determined by energy density and moisture content.

Anthracite, commonly known as "hard coal," is the highest quality of coal, characterized by its exceptional carbon content, which ranges from 80% to 95%. This type of coal has experienced the most extensive coalification process, making it the most efficient and cleanest burning option. Its notable features include a shiny, jet-black appearance and a dense, hard texture, resulting from intense geological forces compressing and heating organic material over millions of years. Anthracite burns at a higher temperature than other coal types, producing minimal smoke and ash, which makes it highly sought after for residential heating, particularly in stoves and fireplaces. Furthermore, its high calorific value and low sulfur content enhance its reputation as a more environmentally friendly coal choice. In industrial settings, anthracite is frequently used in metallurgy, especially for smelting iron and steel, due to its high energy output and clean combustion. However, it is relatively scarce, representing only a small portion of global coal reserves, and its extraction is more difficult due to the depths at which it is typically located. This combination of efficiency, rarity, and specialized uses makes anthracite a premium fuel source in both domestic and industrial applications.

Bituminous coal is a medium-grade coal that lies between anthracite and sub-bituminous types in the coalification process. It is one of the most widely used and readily available forms of coal, making it a significant global energy source. With a carbon content ranging from 60% to 80%, bituminous coal provides a high calorific value and moderate levels of volatile matter. It is mainly used for electricity generation, industrial heating, and as a crucial raw material in coke production, which is essential for the steel industry. Bituminous coal is further divided into thermal coal, used for energy production, and coking coal, valued for its role in metallurgy. In appearance, it is darker and less glossy than anthracite, with a soft, layered structure that allows for easier ignition. However, its combustion releases relatively higher levels of sulfur and nitrogen oxides compared to higher-grade coals, contributing to environmental pollution if not properly managed. Despite these challenges, bituminous coal remains a vital component of energy and industrial processes, especially in regions where alternative energy sources are still developing. Ongoing advancements in clean coal technologies aim to reduce its environmental impact while maximizing its efficiency.

Sub-bituminous coal is a lower grade of coal situated between bituminous coal and lignite in the coalification spectrum. With a carbon content typically ranging from 35% to 60%, it has a relatively lower calorific value compared to bituminous coal but is more efficient and cleaner-burning than lignite. This type of coal is characterized by its higher moisture content, which can range from 20% to 30%, and lower sulfur content, making it a more environmentally favourable option in terms of emissions. Sub-bituminous coal is primarily used for power generation in thermal power plants due to its abundant availability and ease of combustion. Its cleaner-burning properties, compared to other coal types, make it an attractive choice for energy production in regions prioritizing environmental regulations. Sub-bituminous coal typically presents a dull, black or brownish-black appearance and is softer and less dense compared to higher-grade coals. It is commonly found in more recent geological coal deposits. Although it has certain benefits, its high moisture content can create difficulties in transportation and storage, as it might spontaneously combust if not handled properly. This type of coal is vital in the energy strategies of numerous countries, acting as a transitional fuel as they move towards more sustainable and renewable energy sources.

Lignite, also known as "brown coal," represents the lowest rank in the coalification process, with a relatively low carbon content of 25–35% and a high moisture content that can surpass 30%. It forms from partially decomposed organic material under lower heat and pressure than higher-grade coals, making it geologically younger. Due to its high moisture and volatile matter, lignite has the lowest energy content among coal types, leading to less efficiency as a fuel. Nevertheless, it remains an essential energy source, especially in power plants near lignite deposits, as transporting it over long distances is inefficient due to its crumbly nature and tendency to lose energy density during handling. Lignite is soft, brownish-black, and often appears fibrous, reflecting its plant-based origins. Its combustion produces significant ash and emissions, including carbon dioxide, making it less environmentally friendly. However, modern technologies like carbon capture and storage (CCS) are being investigated to reduce its environmental impact. While lignite is important in areas with limited access to higher-grade coals, the global shift towards renewable energy and cleaner fuels may diminish its role in the future. Its use underscores the need to balance energy demands with sustainability goals.

Peat represents the initial stage in the coalification process, consisting of partially decayed organic matter, mainly plant material, in waterlogged and oxygen-poor environments like swamps and bogs. With a carbon content of less than 25%, peat is not classified as coal but is often seen as a precursor due to its potential to transform into lignite under pressure and heat over geological timescales. Peat has a high moisture content, often exceeding 50%, making it a low-energy fuel compared to more developed forms of coal. In its natural state, it is dark brown, fibrous, and spongy, containing visible plant remnants. Despite its low energy value, peat has traditionally been used for heating and cooking in areas where other energy sources are scarce. Beyond energy use, peat is valued in agriculture and horticulture as a soil conditioner due to its high-water retention and organic richness. However, extracting and using peat raises significant environmental concerns, as peatlands are crucial carbon sinks that store large amounts of carbon. Disturbing these ecosystems through peat harvesting or drainage releases stored carbon dioxide and accelerates climate change.

2. Oil

Petroleum, commonly referred to as oil, is a liquid fossil fuel primarily composed of hydrocarbons. Its formation occurs when marine microorganisms' remains settle on the ocean floor and undergo burial beneath sediment layers. These remains transform into liquid hydrocarbons through the application of heat and pressure. Oil extraction is typically achieved through drilling processes and is frequently discovered in conjunction with natural gas deposits.

Procedure of Oil Formation

The oil formation process encompasses several stages. Initially, organic matter, such as plankton, accumulates on the ocean floor, intermingling with sediments. As sediment layers increase, the organic material experiences deeper burial and exposure to heat and pressure, resulting in its transformation into kerogen, a waxy substance. Continued exposure to these conditions causes kerogen to break down, forming crude oil in the form of liquid hydrocarbons. This oil then migrates through porous rock formations, ultimately accumulating in reservoirs. Refineries process crude oil to yield various products, including gasoline, diesel, jet fuel, heating oil, liquefied petroleum gas (LPG), and petrochemicals. Significant oil reserves are primarily located in the Middle East, encompassing countries such as Saudi Arabia, Iraq, and Iran, as well as in Russia, the United States, Canada, and Venezuela. The Organization of the Petroleum Exporting Countries (OPEC) and other major oil-producing nations play a pivotal role in global oil production and pricing mechanisms.

3. Natural Gas

Natural gas, predominantly composed of methane (CH_4), also includes minor quantities of hydrocarbons such as ethane, propane, and butane. This fossil fuel is generated through organic matter decomposition over millions of years, subjected to high pressure and temperature deep within the Earth's crust. Its cleaner combustion properties than coal and oil make natural gas an efficient and environmentally favorable energy source. It is extensively used in electricity generation, heating, and as vehicular fuel in compressed natural gas (CNG) and liquefied natural gas (LNG) forms. Natural gas is also a vital raw material in the petrochemical sector for producing fertilizers, plastics, and chemicals. Its relatively lower carbon dioxide emissions per energy unit make it a transitional fuel as the global energy landscape shifts towards renewables. However, challenges like methane leakage during extraction and transportation pose significant greenhouse gas emission concerns. Extraction methods, particularly hydraulic fracturing (fracking), have raised environmental issues related to water contamination and ecosystem disruption. Despite these challenges, natural gas remains crucial in the global energy mix, with ongoing efforts to mitigate its environmental impact and enhance sustainability. Natural gas, another fossil fuel, consists primarily of methane (CH_4) with minor quantities of other hydrocarbons like ethane, propane, butane, and pentane. It is typically found in deep underground rock formations, often associated with oil deposits. Gas production and extraction involve geologists using seismic surveys and other techniques to identify deposits. Wells are drilled to access these reservoirs, with offshore drilling methods used to extract gas from beneath the ocean floor. Hydraulic fracturing, involving high-pressure fluid injection into rock formations, is used to release trapped natural gas. Raw natural gas is processed to eliminate impurities and separate valuable hydrocarbons before pipeline transportation to consumers.

Figure 13: Fossil Fuels: Coal, Natural Gas, Petroleum

Disadvantages of Conventional energy sources

1. The combustion of fossil fuels emits substantial quantities of carbon dioxide (CO_2) and other greenhouse gases, significantly contributing to global warming and climate change.
2. Contamination of water sources can occur due to oil spills, coal mining activities, and fracking operations, adversely affecting aquatic ecosystems and water quality.
3. Fossil fuel extraction often involves environmentally destructive practices, such as mountaintop removal and offshore drilling, which can severely impact landscapes and ecosystems.
4. The burning of fossil fuels produces pollutants, including sulfur dioxide (SO_2), nitrogen oxides (NOx), and particulate matter, leading to the formation of smog, acid rain, and respiratory health issues.
5. Coal combustion generates significant quantities of fly ash and slag, which may contain toxic heavy metals and require proper disposal methods.
6. The non-renewable nature of fossil fuels implies their eventual depletion, posing a threat to long-term energy security.

Non-Conventional Sources of Energy

Non-conventional energy sources, alternatively referred to as renewable or alternative energy sources, represent environmentally sustainable options that naturally replenish. These sources are gaining prominence as global efforts intensify to mitigate greenhouse gas emissions and reduce dependence on fossil fuels. The primary categories of non-conventional energy sources include:

1. Hydroelectric Energy

This energy form is generated through the conversion of kinetic energy into mechanical energy. Water descends from an elevated position onto a turbine, and the resulting rotation produces electricity. This method of power generation is considered both economical and versatile. Hydropower facilities contribute over 16% of global electricity supply and exhibit minimal environmental impact. Hydropower serves as a significant renewable energy source worldwide, with nations such as China, Brazil, Canada, the United States, and Russia possessing substantial hydroelectric capacity. Moreover, it is regarded as an environmentally friendly energy source. Small-scale hydropower and run-of-river projects are under development to minimize ecological disruption and utilize smaller water bodies. These projects fall into three distinct categories:

1. **Pumped Storage:** This involves two reservoirs at different elevations. During periods of low electricity demand, water is pumped from the lower reservoir to the upper reservoir. During peak demand, water is released back to the lower reservoir to generate electricity.
2. **Dam-based:** This method employs dams to elevate the storage of water in reservoirs. By discharging water from the reservoir to power a turbine connected to a generator, electricity is generated.
3. **Run-of-River:** This approach diverts a portion of the river's flow through a canal or penstock to generate electricity without significantly altering the river's flow or requiring a large reservoir.

Advantages of Hydro Electric Power Plants

1. Hydropower relies on the natural water cycle and is replenished by rainfall, making it a sustainable energy source.
2. Dams can help with flood control and provide a reliable water supply for irrigation and domestic use.
3. Hydropower plants produce very low greenhouse gas emissions compared to fossil fuel-based power plants.
4. Hydropower plants can provide a steady and reliable supply of electricity, especially with large reservoirs.

Figure 14: Dam based Hydroelectric Power Plant

2. Nuclear Energy

Nuclear power plants convert the energy from fission reactions into electricity. This process begins in the reactor core, where a neutron impacts the nucleus of a uranium-235 or plutonium-239 atom, initiating nuclear fission. This reaction splits the atom into two smaller nuclei, releasing substantial heat energy and additional neutrons, which sustain a chain reaction. The heat generated is absorbed by a circulating coolant. In pressurized water reactors, the primary coolant is kept under high pressure to prevent boiling, whereas in boiling water reactors, the coolant vaporizes directly within the core. The heated coolant either transfers its thermal energy to water in a steam generator (pressurized water reactor) or generates steam directly (boiling water reactor). The produced steam drives a turbine connected to an electric generator. Through electromagnetic induction, the turbine's rotational energy is converted into electrical energy. After passing through the turbine, the steam is cooled and condensed in a condenser, facilitated by cooling water from a nearby source or cooling towers. The condensed water is then recirculated to the steam generator, continuing the cycle.

Advantages of Nuclear Energy

1. Nuclear fission releases a substantial amount of energy compared to chemical reactions, allowing a small quantity of nuclear fuel to produce a large amount of electricity.
2. Nuclear power plants provide a stable and uninterrupted electricity supply, making them suitable for base-load power generation.
3. Uranium, the primary fuel for nuclear reactors, is relatively abundant and can be sourced from various global regions.

Disadvantages of Nuclear Energy

1. The technology and materials used in nuclear power production can potentially be diverted for nuclear weapons manufacturing, posing security risks. High-level radioactive waste requires secure storage for thousands of years. Deep geological repositories are under development for long-term storage solutions.
2. Environmental exposure to nuclear materials can cause significant harm to living organisms.
3. The construction of nuclear power plants involves substantial capital investment, lengthy regulatory processes, and complex engineering challenges.

Figure 15: Nuclear Power Environmental Collage

3. Solar Energy

Solar energy, derived from solar radiation, offers an alternative method of electricity generation. This renewable resource is particularly valuable during periods of high demand, such as hot days when air conditioning usage peaks. Solar power production is characterized by its absence of operational emissions. The proliferation of solar installations reduces the necessity for polluting power plants. In the United States, the majority of the two million solar installations are grid-connected, providing users with credits that offset electricity costs. Solar power contributes to states' compliance with Clean Energy Standards, diminishing reliance on more costly and environmentally detrimental peak generating units. Beyond conventional photovoltaic systems, solar technologies are being integrated into other applications. Ongoing development focuses on concentrating solar technologies for distributed heat and power systems, capable of providing localized electricity and thermal energy without emissions. Two primary solar energy technologies exist: one employs large mirror fields to collect and convert solar energy into heat, while the other utilizes photovoltaic cells for direct electricity generation. The former requires approximately 10 acres of sun-exposed land for a typical plant. These heat-based systems, which drive electricity-generating turbines, have achieved commercial viability and have been operational for over a decade in some instances.

The thermal energy generated can be harnessed to power various machines that produce electricity, transform liquids or gases into steam, or provide direct industrial heating systems utilized in the manufacture of ethanol, cement, or steel. Concentrating solar heat systems can be employed to deliver electricity or hot water, particularly in remote rural regions or areas with high electricity costs or where pollution reduction is a priority. Small-scale, distributed concentrating solar heat systems integrated into manufacturing facilities can drive the production of electricity, hydrogen, ethanol, and cement. Recent advancements in solar thermal power technology have sparked debates regarding scalable methods for providing sustainable and renewable energy to urban consumers.

Solar Energy Classifications

1. Photovoltaic (PV) Solar Energy

Photovoltaic (PV) solar energy entails the direct conversion of sunlight into electricity via photovoltaic cells, typically made from semiconductor materials like silicon. When sunlight interacts with these cells, photons excite electrons within the semiconductor, generating an electric current—a phenomenon known as the photovoltaic effect. This current is harnessed and transformed into usable electricity through inverters. PV systems are adaptable and scalable, ranging from small residential rooftop panels to expansive solar farms powering entire communities. The benefits of photovoltaic solar energy include its capacity to generate clean, renewable electricity without greenhouse gas emissions during operation, thereby aiding in climate change mitigation. Additionally, PV systems can be implemented in both urban and remote locations, making them suitable for decentralized energy production. However, challenges such as efficiency losses due to weather conditions, land requirements for large-scale installations, and the environmental impact of solar panel manufacturing persist. Despite these obstacles, advancements in PV technology, including high-efficiency cells and bifacial panels, continue to enhance performance, reliability, and cost-effectiveness, establishing photovoltaic solar energy as a pivotal component in the global energy transition.

2. Concentrated Solar Power (CSP)

CSP systems utilize solar radiation to generate thermal energy, subsequently converted into electricity. In contrast to photovoltaic panels, CSP employs mirrors or lenses to concentrate substantial amounts of sunlight onto a designated area, typically a receiver containing a heat transfer medium such as molten salts, synthetic oil, or water. The concentrated heat elevates the temperature of the medium, which is then utilized to produce steam that drives turbines connected to generators, ultimately resulting in electricity generation. CSP systems are particularly advantageous for utility-scale applications due to their capacity to store thermal energy for use during overcast conditions or nighttime, thereby ensuring a consistent and reliable power output. Common CSP configurations include parabolic trough systems, solar towers, linear Fresnel systems, and dish/engine systems, each offering distinct advantages and efficiencies contingent upon geographic and technical requirements. Despite its potential, CSP encounters challenges such as high initial costs, land usage requirements, and the necessity for abundant sunlight, which limits its applicability to regions with high solar radiation. Nevertheless, advancements in CSP technologies, including improved thermal storage solutions and hybrid systems integrating other renewable sources, are enhancing its economic viability and environmental benefits, rendering it an increasingly attractive option for sustainable energy generation.

3. Passive Solar Energy

Passive solar energy refers to the method of utilizing sunlight to naturally provide heating, cooling, and lighting without employing mechanical systems or active energy conversion processes. This approach emphasizes architectural design and building orientation to optimize the absorption and distribution of solar energy within a structure. Key strategies include the use of large south-facing windows to capture maximum sunlight during winter months, thermal mass materials such as concrete or brick to store and release heat, and effective insulation to prevent heat loss. Passive solar energy also incorporates features such as shading devices, overhangs, and reflective surfaces to regulate indoor temperatures during summer, preventing overheating. Unlike active solar systems that employ photovoltaic panels or solar thermal devices, passive solar energy relies entirely on natural energy flow, reducing costs and maintenance requirements. It plays a significant role in sustainable building designs by minimizing reliance on artificial heating and cooling systems, thereby lowering energy consumption and greenhouse gas emissions. Passive solar techniques can be integrated into new constructions and retrofitted into existing buildings, offering an efficient and environmentally friendly solution to maximize energy savings while enhancing indoor comfort. This method exemplifies how thoughtful design can leverage renewable energy in simple yet impactful ways.

4. Solar Thermal Energy

Solar thermal energy involves the utilization of sunlight to generate heat, which can be used directly or converted into other forms of energy such as electricity. Unlike photovoltaic systems that convert sunlight into electricity, solar thermal systems capture and concentrate solar radiation to heat a fluid, which in turn transfers the heat for various applications. Solar thermal systems can be categorized into three primary types based on their temperature ranges: low-temperature systems for tasks like solar water heating, medium-temperature systems for heating air or industrial processes, and high-temperature systems for power generation, such as those used in concentrated solar power (CSP) plants. These systems often employ collectors like flat plates, evacuated tubes, or parabolic mirrors to optimize sunlight absorption. Solar thermal energy is extensively used in residential, commercial, and industrial environments, especially for water heating, space heating, and process heating. One of its benefits is the reduction in dependence on fossil fuels, which leads to lower carbon emissions and energy expenses. However, challenges such as efficiency losses in areas with low sunlight and the requirement for large-scale infrastructure for CSP systems can restrict its use. Progress in storage technologies and hybrid systems is helping to overcome these obstacles, making solar thermal energy a crucial part of the global movement towards sustainable energy solutions.

5. Solar Desalination

Solar desalination energy harnesses solar power to extract salt and impurities from seawater or brackish water, rendering it suitable for drinking, agriculture, and industrial purposes. This eco-friendly process utilizes the Sun's abundant and renewable energy, eliminating the need for fossil fuels and reducing greenhouse gas emissions. Solar desalination can be executed through two main approaches: direct and indirect systems. Direct solar desalination uses solar stills, which replicate the natural water cycle by evaporating water with solar heat and then condensing the vapor into fresh water. In indirect systems, solar energy powers technologies like reverse osmosis or multi-effect distillation, which are more efficient for large-scale operations. Solar desalination is especially beneficial in arid and remote areas where access to freshwater is limited but sunlight is abundant. The advantages of this technology include minimal environmental impact, low operating costs, and the potential to decentralize freshwater production. However, challenges such as high initial costs, the need for technological optimization, and slower output compared to conventional desalination methods can hinder widespread adoption. Ongoing improvements in solar energy efficiency and cost reduction are making solar desalination an increasingly feasible solution for tackling global water scarcity.

6. Solar Cooking

Solar cooking energy uses the Sun's radiant heat to cook food, boil water, or pasteurize liquids without relying on traditional fuels like wood, coal, or gas. It utilizes solar cookers, which are specially designed devices that capture, concentrate, and retain sunlight to generate the necessary heat for cooking. There are three main types of solar cookers: box cookers, parabolic cookers, and panel cookers. Box cookers use reflective panels and a transparent cover to trap sunlight, heating food placed inside. Parabolic cookers focus sunlight onto a single point, achieving high temperatures suitable for frying or grilling, while panel cookers combine elements of reflectors and insulated surfaces for simpler cooking tasks. The benefits of solar cooking include its eco-friendly nature, as it eliminates smoke and greenhouse gas emissions, and its affordability, requiring only sunlight as an energy source. Solar cooking offers significant advantages in remote areas where fuel access is restricted, fostering energy autonomy and diminishing reliance on deforestation for firewood. Nevertheless, its effectiveness is highly contingent upon weather conditions, necessitating direct and consistent sunlight. Recent advancements in solar cooker designs are addressing these limitations, rendering solar cooking a viable and sustainable solution for both household and community applications. This method not only tackles energy and health challenges but also contributes to environmental conservation efforts.

Advantages of Solar Energy

1. Absence of pollution during operation.
2. Serves as an effective complement to other renewable energy sources.
3. Enables the operation of electronic devices in areas lacking electrical infrastructure.

Disadvantages of Solar Energy

1. Preferential implementation in regions with high solar irradiance.
2. Reliance on the availability of sunlight.
3. Substantial installation costs associated with large-scale solar panel arrays.

Figure 16: Solar Panels

Biomass

The reliance on fossil fuels, such as coal, petroleum, and natural gas, has led to numerous environmental challenges. The extraction, transport, and utilization of these resources not only raise ecological concerns but also contribute to regional and international political tensions. Therefore, it is crucial to diversify the energy supply by incorporating renewable energy sources. Currently, technologies harnessing alternative energy resources, such as hydropower, wind, solar, and nuclear energy, which exhibit reduced environmental impacts, are being introduced. Moreover, as the establishment of a sustainable society requires more than just electricity and heat generation, it is imperative to actively employ energy derived from natural resources for heating, cooling, gas heat production, and various industrial applications. Biomass energy resources, which can be perpetually utilized through the cyclical process of photosynthesis, play a significant role in addressing contemporary environmental and energy issues. Biomass energy, a renewable energy form derived from organic matter such as plant and animal waste, can be converted into heat, electricity, or fuels. Its sustainability stems from its ability to recycle atmospheric carbon absorbed by plants during photosynthesis. Notable biomass conversion technologies include combustion, pyrolysis, gasification, and fermentation.

Biomass Production

The production of biomass energy entails the transformation of organic materials into usable energy forms, beginning with the cultivation or collection of biomass feedstocks. These feedstocks encompass agricultural residues (e.g., straw, husks, and manure), forestry byproducts (e.g., wood chips and sawdust), purpose-grown energy crops (e.g., switchgrass, willow, and miscanthus), and organic waste from urban and industrial sources. Following collection, the biomass undergoes various conversion processes tailored to the desired end product. Direct combustion of biomass yields heat and electricity, while gasification converts biomass into syngas through partial combustion in an oxygen-limited environment. Pyrolysis decomposes biomass at high temperatures in the absence of oxygen, producing bio-oil, syngas, and biochar. Anaerobic digestion employs microorganisms to break down organic matter in oxygen-free conditions, generating biogas and nutrient-rich digestate. Fermentation utilizes microorganisms such as yeast to convert biomass sugars into ethanol. These conversion processes transform biomass into various energy forms, including electricity, heat, transportation fuels, and biogas, thereby contributing to renewable energy production and waste management strategies.

Benefits of Biomass Energy Sources

1. Reduces reliance on fossil fuels and promotes their conservation.
2. Considered carbon-neutral due to the balance between CO_2 released during combustion and CO_2 absorbed by plants during growth.
3. Renewable through natural processes, ensuring long-term sustainability.
4. Stimulates agricultural and forestry sectors, generating employment opportunities and supporting rural economies.
5. Offers a versatile and sustainable renewable energy source that contributes to greenhouse gas reduction, waste management, and rural economic development.

Figure 17: Biomass Processing

Biogas

Anaerobic conditions in specialized reactors facilitate the production of biogas, a renewable energy source derived from the biodegradation of organic matter. This gaseous mixture primarily comprises methane and carbon dioxide, with methane concentrations typically ranging from 40% to 80%. The versatility of biogas is evident in its diverse applications, including electricity generation, heating, fuel production, and use as a raw material. Notably, biogas production from waste serves as a crucial waste management strategy, addressing multiple environmental concerns such as air and water pollution mitigation, and reduction of atmospheric methane emissions. This approach is particularly beneficial in urban environments with spatial constraints for waste management. Among renewable energy technologies, biogas holds a prominent position due to its practicality and wide-ranging applications. While extensive research has been conducted on waste-derived biogas production, it is important to acknowledge that despite its advantages, this process presents significant environmental challenges. The comprehensive utilization of organic matter for biogas generation results in residual, non-beneficial liquor. Various sources contribute to biogas production, including agricultural waste, sewage sludge, energy crops, and municipal and industrial waste.

Biogas Production Process

The production of biogas involves the anaerobic digestion of organic materials by microorganisms in oxygen-deprived environments. This process commences with the accumulation of biomass feedstocks, which may encompass agricultural residues, municipal organic waste, sewage sludge from wastewater treatment facilities, and industrial organic byproducts. The collected biomass is then introduced into a sealed, anaerobic digester. Within this environment, a sequence of microbial processes unfolds: hydrolysis initially breaks down complex organic compounds into simpler molecules; acidogenesis subsequently converts these molecules into volatile fatty acids and alcohols; acetogenesis then transforms these intermediates into acetic acid, hydrogen, and carbon dioxide; finally, methanogenesis yields biogas, predominantly composed of methane and carbon dioxide. The captured biogas can be utilized for various purposes, including electricity and heat generation, cooking, or further refined into biomethane for vehicular fuel. The remaining digestate serves as a nutrient-rich byproduct, suitable for use as fertilizer, thereby enhancing soil quality and promoting nutrient cycling in agricultural ecosystems.

Limitations of Biogas Plants

1. Establishing biogas facilities requires substantial initial capital for digesters, storage, and purification systems.
2. Continuous biogas production demands a reliable supply of organic waste feedstock, which may vary seasonally.
3. Impurities in biogas, such as hydrogen sulfide (H_2S), require purification, increasing production costs and complexity.
4. Anaerobic digestion can produce malodorous emissions, raising concerns in neighbouring communities if not properly managed.
5. Cultivating energy crops for biogas may compete with food production and affect land availability, leading to land-use conflicts. Additionally, biogas energy yield is often lower than that of other renewable sources like solar or wind, making it less efficient in some cases.

Hydrogen as an Alternative Future Source of Energy

The escalating energy demand coupled with dwindling fossil fuel reserves has prompted extensive research into novel, clean energy alternatives. Hydrogen emerges as a particularly promising candidate for future energy needs, attributed to its high-octane power, content, reaction enthalpy, regenerative properties, absence of greenhouse gas emissions, non-polluting nature, and widespread availability. Various natural and alternative sources, including wind, solar, hydro, biomass, and nuclear power, can be harnessed for hydrogen production. Practical applications of hydrogen technology encompass fuel cells for personal vehicles and electrical power plants, offering potential benefits such as clean energy generation, reduced grid pollution, and advancements in transportation and fuel production. Evidence suggests that once a competitive price point is achieved, hydrogen adoption may accelerate more rapidly than anticipated, with utilization rates closely correlated to oil price fluctuations. Current global hydrogen production, predominantly through natural gas reforming, stands at approximately 50 million tons. However, the widespread adoption of hydrogen technology faces challenges, including high dispensing costs and the need for enhanced energy efficiency. Given the current economic constraints, government support plays a crucial role in sustaining hydrogen usage over extended periods. Significant efforts are underway to bolster hydrogen research infrastructure, including R&D initiatives, collaborative projects with industry laboratories, and material project promotion. Concurrently, industrial entities have implemented adaptation strategies that have contributed to recent reductions in production and utilization costs. These policy implementations indicate a projected demand for hydrogen reaching 50-100 billion by 2020 and 100-200 billion by 2030, necessitating the development of new large-scale energy production facilities to meet this unprecedented demand.

Hydrogen Production Methods can be Categorized as Follows

1. **Grey Hydrogen:** Grey hydrogen is defined as hydrogen produced from natural gas through a process known as steam methane reforming (SMR), which is currently the most prevalent and cost-effective method of hydrogen production. In this process, natural gas is heated with steam in the presence of a catalyst to yield hydrogen and carbon dioxide as by-products. Although grey hydrogen is a crucial component of the global hydrogen supply chain, it is not deemed environmentally friendly due to its substantial greenhouse gas emissions, primarily from the CO_2 released during production. Grey hydrogen serves as the baseline in hydrogen production, being relatively low-cost and technologically well-established, thus making it a popular choice for industries such as ammonia production, refining, and chemical manufacturing. However, its dependence on fossil fuels undermines its sustainability and contributes to climate change, prompting increased interest in low-carbon alternatives such as green and blue hydrogen. Transitioning from grey hydrogen to cleaner methods remains challenging due to the infrastructure and economic investments required. Nonetheless, grey hydrogen continues to play a role in meeting global energy demands while providing a foundation for advancing cleaner hydrogen technologies. Efforts to integrate carbon capture and storage (CCS) into grey hydrogen production could help mitigate its environmental impact, evolving it into "blue hydrogen."
2. **Blue Hydrogen:** Blue hydrogen refers to hydrogen produced through the steam methane reforming (SMR) process, akin to grey hydrogen, but with the critical addition of carbon capture, utilization, and storage (CCUS) technology to mitigate environmental impact. In this process, natural gas is heated with steam in the presence of a catalyst to produce hydrogen and carbon dioxide. The key distinction of blue hydrogen is that a significant portion of the resulting CO_2 is captured and either stored underground or repurposed for industrial uses, thereby reducing greenhouse gas emissions. This positions blue hydrogen as a low-carbon

energy source, serving as a bridge between conventional fossil fuel-based hydrogen production and the completely renewable green hydrogen. Blue hydrogen is considered particularly valuable for decarbonizing hard-to-abate industries such as steel, cement, and chemicals, as well as for use in power generation and transportation. While blue hydrogen provides a cleaner alternative to grey hydrogen, it still relies on natural gas, raising concerns about methane leakage during extraction and transportation. Additionally, the cost and scalability of CCUS technology can pose challenges for its widespread adoption. Despite these limitations, blue hydrogen is seen as a critical transitional solution in the global effort to achieve net-zero emissions, enabling progress toward a sustainable hydrogen economy.

3. **Green Hydrogen:** Green hydrogen is a clean energy carrier produced through electrolysis, where water is split into hydrogen and oxygen using electricity from renewable sources like wind, solar, or hydropower. This method ensures no carbon emissions during production, making green hydrogen essential for global decarbonization and mitigating climate change. Green hydrogen's versatility allows its use across energy storage, transportation, industry, and power generation sectors. It stores surplus renewable energy, powers fuel-cell electric vehicles (FCEVs), replaces fossil fuels in industrial processes, and produces chemicals sustainably.

   Despite its potential, green hydrogen faces challenges. Production remains costly due to expenses associated with electrolyzes and renewable energy infrastructure. Efficiency losses during production, storage, and utilization pose technical barriers. Infrastructure for hydrogen storage, transportation, and distribution is limited, and electrolysis' water-intensive nature could be problematic in water-scarce regions. However, ongoing research and international collaboration are advancing green hydrogen technology to reduce costs, improve efficiency, and expand infrastructure. Countries like Germany, Japan, India, and Australia are investing substantially and developing strategic roadmaps to foster a global hydrogen economy. Green hydrogen represents a promising avenue for achieving carbon neutrality and addressing environmental sustainability, making it critical for future energy systems.

4. **Pink Hydrogen:** Pink hydrogen is a variant of hydrogen produced through electrolysis powered specifically by nuclear energy. This method utilizes electricity generated from nuclear power plants to dissociate water molecules into hydrogen and oxygen. The distinguishing

characteristic of pink hydrogen is its dependence on nuclear energy, which is regarded as low-carbon and consistent, rendering it a dependable option for hydrogen production. Given that nuclear power is not contingent on weather conditions like solar or wind, pink hydrogen provides a stable and scalable solution for producing clean hydrogen without greenhouse gas emissions during production. Additionally, it benefits from the high energy density of nuclear fuel, which facilitates efficient hydrogen generation. Pink hydrogen is garnering interest as a viable alternative to other forms of hydrogen, particularly in regions with established nuclear infrastructure. It is also perceived as a means to optimize the use of nuclear plants, especially during periods of surplus electricity generation. Despite these advantages, the adoption of pink hydrogen faces challenges, such as public concerns regarding nuclear safety, waste disposal, and the high costs associated with constructing and maintaining nuclear reactors. Nonetheless, pink hydrogen holds potential as a transitional energy source that amalgamates the clean attributes of hydrogen with the reliability and scalability of nuclear power, contributing to the broader objective of achieving carbon neutrality.

Limitations of Hydrogen Technology

Despite its potential, hydrogen technology faces certain limitations:

1. The production cost of green hydrogen currently exceeds that of fossil fuel-based hydrogen.
2. The development of comprehensive infrastructure for hydrogen production, storage, and distribution presents a significant challenge.

Very short Answer Type Questions

1. What is the primary source of global energy consumption?
2. Name three types of fossil fuels.
3. What is the highest rank of coal?
4. What are two primary solar energy technologies?
5. What is biogas primarily composed of?
6. What are the three main types of hydrogen production?
7. Discuss Photovoltaic Solar Energy.
8. Discuss the advantages of solar energy.
9. Tell the steps involved in biogas formation process.
10. What is the use of green hydrogen?

Short Answer Type Questions

1. Briefly explain the process of coal formation.
2. List three advantages and three disadvantages of hydroelectric power.
3. Describe two methods of solar energy production.
4. Explain the difference between grey, blue, and green hydrogen.
5. What are two limitations of biogas plants.
6. Discuss two hydrogen production methods.
7. Discuss the benefits of biomass energy.
8. Discuss the pros and cons of solar energy.
9. Explain P ink hydrogen

Long Answer Type Questions

1. Discuss the environmental impacts of the Industrial Revolution's shift to fossil fuels.
2. Compare and contrast renewable and non-renewable energy sources, providing examples of each.
3. Explain the process of coal formation and its different ranks.
4. Describe the process of biogas production and its advantages and limitations.
5. Discuss hydroelectric energy and also tell its advantages over conventional sources.
6. Differentiate between Conventional and Non-Conventional Sources with examples.
7. What is Electromagnetic Radiation? Describe their applications in daily use.
8. Discuss solar energy in detail. Also mention all its categories and applications.

Multiple Choice Questions

1. The Industrial Revolution primarily shifted energy production from:

   a) Fossil fuels to renewable sources
   b) Manual labor and animals to fossil fuels
   c) Nuclear energy to solar energy
   d) Hydropower to wind power

2. Which of the following is NOT a primary energy source?

   a) Solar energy
   b) Electricity
   c) Natural gas
   d) Geothermal energy

3. Which type of coal has the highest carbon content and energy density?

   a) Lignite
   b) Bituminous coal
   c) Sub-bituminous coal
   d) Anthracite

4. The primary constituent of natural gas is:

a) Ethane
b) Propane
c) Butane
d) Methane

5. Which of the following is a disadvantage of fossil fuels?
   a) High energy density
   b) Relatively inexpensive
   c) Contribution to climate change
   d) Reliable energy supply
6. Which type of hydroelectric power plant diverts a portion of a river's flow without significantly altering it?
   a) Pumped storage
   b) Dam-based
   c) Run-of-river
   d) Tidal
7. What is the primary fuel used in nuclear reactors?
   a) Plutonium-239 only
   b) Uranium-235 only
   c) Uranium-235 and Plutonium-239
   d) None of the above
8. Biomass energy is considered carbon-neutral because:
   a) It produces no greenhouse gases.
   b) It uses only waste materials.
   c) CO2 released during combustion is counterbalanced by the CO_2 absorbed during plant growth.
   d) It is a very efficient energy source.
9. What is the main component of biogas?
   a) CO_2
   b) CH_4
   c) H
   d) O

7

Ecosystem Services and Valuation

Introduction

Ecosystems are the foundation of life on Earth, providing a vast array of benefits that are essential for human well-being, economic prosperity, and environmental health. These benefits, often referred to as ecosystem services, are the direct and indirect contributions of natural systems to human societies. Understanding and appreciating these services is crucial for sustainable development and the preservation of biodiversity.

Provisioning Services

Provisioning services are the tangible products that ecosystems supply. These include food, fresh water, timber, fiber, fuel, and medicinal resources. Forests, for example, provide wood for construction and fuel, as well as non-timber products like fruits, nuts, and resins. Oceans, rivers, and lakes offer fish and seafood, which are vital sources of protein for millions of people. Plants and animals are sources of traditional and modern medicines, and genetic resources from wild species contribute to crop improvement and pharmaceutical development.

Water is another critical provisioning service. Natural ecosystems like forests, wetlands, and watersheds regulate the flow and quality of water, ensuring a steady supply for drinking, agriculture, and industry. The availability of clean water is fundamental to public health and economic activities.

Regulating Services

Regulating services are the benefits that ecosystems provide by maintaining environmental conditions. These include climate regulation, air and water purification, flood control, disease regulation, and pollination. Forests and oceans act as carbon sinks, absorbing carbon dioxide and helping to mitigate climate change. Wetlands filter pollutants from water, reducing the need for expensive water treatment infrastructure.

Pollination by insects, birds, and other animals is essential for the reproduction of many crops and wild plants. Without pollinators, food production would decline dramatically, threatening food security. Natural predators and healthy ecosystems also regulate pests and diseases, reducing the reliance on chemical pesticides and supporting sustainable agriculture.

Flood control is another vital regulating service. Wetlands, mangroves, and floodplains absorb excess water during storms and heavy rains, reducing the risk of floods and protecting communities and infrastructure.

Cultural Services

Cultural services are the non-material benefits that people derive from ecosystems. These include recreation, tourism, aesthetic enjoyment, spiritual fulfillment, and educational opportunities. Natural landscapes such as mountains, forests, rivers, and beaches provide spaces for outdoor activities like hiking, camping, birdwatching, and water sports. These activities contribute to physical and mental well-being, foster social connections, and support local economies through tourism.

Ecosystems also inspire art, literature, and cultural traditions, shaping identities and worldviews. Sacred groves, rivers, and mountains hold spiritual significance for many communities, and access to nature is linked to improved health and quality of life.

Supporting Services

Supporting services are the underlying processes that make all other ecosystem services possible. These include nutrient cycling, soil formation, primary production, and the maintenance of biodiversity. Healthy soils, formed through the decomposition of organic matter and the activity of soil organisms, are essential for agriculture and natural vegetation.

Nutrient cycling ensures the availability of essential elements like nitrogen and phosphorus, supporting plant growth and food production. Biodiversity, the variety of life forms within ecosystems, underpins resilience and adaptability, enabling ecosystems to recover from disturbances and continue providing services.

The Value of Ecosystem Services

The value of ecosystem services is immense, often surpassing the economic output of traditional industries. However, many of these services are undervalued or ignored in decision-making processes, leading to their degradation and loss.

Deforestation, pollution, habitat destruction, and overexploitation threaten the ability of ecosystems to provide these essential benefits.

Recognizing the true value of ecosystem services is critical for sustainable development. By integrating ecosystem services into economic and policy decisions, societies can make informed choices that balance development with conservation. This approach encourages investments in ecosystem restoration, the creation of protected areas, and the adoption of sustainable practices that safeguard the natural systems on which we depend.

Challenges and the Need for Conservation

Despite their importance, ecosystem services face numerous challenges. Climate change, biodiversity loss, land use change, and pollution are undermining the capacity of ecosystems to function effectively. The loss of pollinators, for example, threatens global food production, while the destruction of wetlands increases the risk of floods and water pollution.

Conserving and restoring ecosystems is essential for maintaining the flow of ecosystem services. This requires coordinated efforts at local, national, and global levels, involving governments, communities, businesses, and individuals. Sustainable management of natural resources, habitat protection, and the promotion of biodiversity are key strategies for ensuring that ecosystems continue to provide benefits for present and future generations.

Resource Management

Resource management is the strategic and sustainable use of natural resources to meet human needs while preserving the health of ecosystems for future generations. It encompasses the planning, allocation, and monitoring of resources such as land, water, soil, plants, and animals, integrating ecological, economic, and social considerations.

Principles of Resource Management

Effective resource management is guided by several core principles:

1. **Sustainability:** The primary objective is to use resources in a way that does not compromise their availability for future generations. This involves maintaining ecosystem health, productivity, and resilience.
2. **Ecosystem Approach:** Modern resource management adopts a holistic perspective, recognizing the interconnectedness of different ecosystem components and the influence of human activities.

3. **Adaptive Management:** Resource management is an ongoing process that requires flexibility and the ability to respond to new information and changing conditions. Adaptive management involves monitoring outcomes, learning from experience, and adjusting strategies as needed.
4. **Stakeholder Participation:** Involving all stakeholders, including local communities, governments, industries, and non-governmental organizations, ensures that diverse perspectives and knowledge systems are integrated into decision-making.

Tools and Strategies

Resource management employs a variety of tools and strategies to achieve its goals:

1. **Regulatory Measures:** Governments establish laws and regulations to control resource use, protect endangered species, and prevent environmental degradation. Examples include fishing quotas, logging restrictions, and protected area designations.
2. **Market-based Instruments:** Economic incentives such as taxes, subsidies, tradable permits, and payment for ecosystem services schemes encourage sustainable practices and internalize the costs of environmental damage.
3. **Restoration and Conservation:** Active efforts to restore degraded ecosystems and conserve biodiversity are central to resource management. This can involve reforestation, wetland restoration, invasive species control, and habitat protection.
4. **Education and Capacity Building:** Raising awareness and building capacity among resource users is vital. Extension services, training programs, and public education campaigns promote best practices and foster stewardship.

Integrating Ecosystem Services

Incorporating ecosystem services into resource management decisions enhances efficiency, reveals trade-offs, and identifies win-win solutions for conservation and development. By explicitly recognizing the benefits that healthy ecosystems provide, managers can prioritize actions that maintain or enhance these services.

For example, protecting forests not only conserves biodiversity but also ensures clean water supplies, sequesters carbon, and supports livelihoods. Restoring wetlands can reduce flood risks, improve water quality, and provide habitat for wildlife. Such integrated approaches often require coordination across sectors, jurisdictions, and scales.

Adaptive and Ecosystem-Based Management

Adaptive management is a dynamic approach that acknowledges uncertainty and the need for learning. By monitoring outcomes and adjusting strategies, managers can respond to changing conditions and improve resource use over time. Ecosystem-based management goes further by considering the cumulative impacts of different activities on entire ecosystems, promoting resilience and long-term sustainability.

Challenges and Future Directions

Resource management faces significant challenges, including population growth, climate change, habitat loss, and competing demands for land and water. Addressing these challenges requires innovative policies, cross-sectoral collaboration, and a commitment to long-term sustainability.

The future of resource management lies in embracing adaptive, ecosystem-based approaches that recognize the interconnectedness of human and natural systems. By valuing and managing ecosystem services wisely, societies can secure the benefits of nature for generations to come.

Conclusion

Resource management is essential for balancing human needs with environmental protection. By applying sound principles, employing effective tools, and integrating ecosystem services into decision-making, we can ensure the sustainable use of natural resources and the continued provision of benefits that support human well-being and ecological health.

1. Landscape ecology studies how spatial arrangement and diversity of ecosystems affect ecological processes.
2. It focuses on **heterogeneity**, examining variations in ecosystem types, shapes, sizes, and distributions.
3. Core elements:
   a. **Composition** – Types and abundance of ecosystems.
   b. **Configuration** – Spatial arrangement (patches, corridors).
   c. **Connectivity** – Movement of species, energy, and materials across the landscape.
4. **Fragmentation** leads to isolated habitat patches, impacting species survival and ecosystem functions.

5. The **matrix** is the dominant land use surrounding habitat patches, influencing connectivity.
6. **Scale** is vital—ecological patterns and processes change depending on spatial and temporal scales.
7. The field integrates **ecology, geography, planning, and social sciences,** recognizing human influence on landscapes.
8. Developed from Carl Troll's work in 1939 using aerial photography.
9. It has applications in **conservation, habitat restoration, land-use planning, and sustainability**.
10. Landscape ecology bridges natural and human systems to manage biodiversity and ecosystem services effectively.

Very Short Answer Type Questions

1. What does landscape ecology study?
2. Who introduced the term "landscape ecology"?
3. Define spatial heterogeneity.
4. What is meant by landscape composition?
5. What is a "matrix" in landscape ecology?
6. Mention one application of landscape ecology.
7. What is habitat fragmentation?
8. Name one discipline integrated in landscape ecology.
9. What is the main focus of configuration in a landscape?
10. What does "connectivity" refer to?

Short Answer Type Questions (2–3 marks)

1. Explain the concept of spatial heterogeneity in landscape ecology.
2. What roles do corridors play in maintaining landscape connectivity?
3. Differentiate between composition and configuration.
4. Describe the role of scale in landscape ecology.
5. How does habitat fragmentation affect species movement?
6. What is the historical origin of landscape ecology?
7. How does human activity influence modern landscapes?
8. Explain the significance of the matrix in determining ecological flows.
9. Why is landscape ecology considered interdisciplinary?
10. List any two major research themes in landscape ecology.

Long Answer Type Questions

1. Discuss the major concepts of landscape ecology: composition, configuration, and connectivity.
2. Describe the effects of fragmentation and the role of the matrix in landscape dynamics.
3. Elaborate on the importance of scale and pattern in landscape ecology.
4. Explain how landscape ecology integrates natural and social sciences.
5. Analyze the influence of spatial heterogeneity on ecological processes.
6. Discuss the relevance of landscape ecology in conservation and habitat restoration.
7. Explain how landscape ecology addresses challenges from human-induced changes.
8. Illustrate the relationship between spatial patterns and ecological functions with examples.
9. Describe the application of landscape ecology in planning and sustainable land management.
10. Summarize the evolution and current relevance of landscape ecology.

Multiple Choice Questions

1. Who coined the term "landscape ecology"?
   a. Eugene Odum **b. Carl Troll**
   c. Rachel Carson d. Charles Elton
2. Which of the following best describes "composition" in landscape ecology?
   a. Land ownership pattern b. Spatial arrangement of roads
   c. Variety and proportion of land covers d. Movement of animals
3. Fragmentation leads to
   a. Increased biodiversity b. Formation of large continuous habitats
   c. Smaller, isolated habitat patches d. Higher soil fertility
4. The matrix in landscape ecology is:
   a. A wildlife corridor b. The least important habitat
   c. The dominant surrounding land cover d. Always a forested area
5. Which of the following is a core concept in landscape ecology?
   a. Succession b. Homeostasis
   c. Connectivity d. Respiration

6. Landscape ecology differs from traditional ecology by emphasizing:
    a. Individual species
    b. Spatial patterns and heterogeneity
    c. Cellular respiration
    d. Genetic variation only
7. Which factor determines how easily organisms move across a landscape?
    a. Trophic level
    b. Connectivity
    c. Biomass
    d. Carbon level
8. A corridor in a landscape helps in
    a. Preventing soil erosion
    b. Enhancing photosynthesis
    c. Facilitating species movement
    d. Slowing river flow
9. A key application of landscape ecology is in
    a. Astronomy
    b. Agricultural trade
    c. Land-use planning
    d. Vaccine development
10. What is the role of "configuration" in a landscape?
    A. Soil formation
    B. Spatial arrangement of elements
    C. Climate regulation
    D. Ocean circulation

Section IV
Pollution, Waste and Environmental Health

8

Environmental Pollution

The growth in population leads to an increased demand for essential resources such as electricity, vehicles, water, and infrastructure. Additionally, it fuels the desire for luxury items like high-end cars, cutting-edge electronic devices including smartwatches and mobile phones, 5G technology, fashionable clothing, powerful sound systems, and multi-story buildings. These human desires places a significant burden on the planet. The escalating human appetite for resources exerts immense pressure on the environment, resulting in its overexploitation. Humans are the primary contributors to various forms of pollution, including air, water, land, and noise pollution, either directly or indirectly. Reckless human activities cause a wide range of pollution types, severely impacting natural resources. This chapter will explore different pollution categories, their effects on public health, and strategies for disaster management.

Human actions are depleting nature, which in turn has detrimental effects on human health. This chapter will examine how human activities impact the environment and its consequent repercussions on human well-being.

Types of Pollution

1. Water Pollution

Water pollution is the primary reason for the scarcity of clean water on Earth. Human negligence has severely impacted water bodies. Unsustainable use and carelessly wastage of water resources, along with the pollution of water bodies, contribute to extensive water contamination. These harmful practices affecting lakes, rivers, ponds, and estuaries gradually diminish the availability of fresh, potable water, leading to the emergence of various water-borne and water-induced diseases, as discussed in detail in chapter 5. Even the Ganga, a river revered for its purity, is now contaminated with numerous waste products. Religious practices, such as the immersion of deity idols and worship items, are a significant cause of its pollution. Several factors contribute to water pollution:

1. **Industrial effluents:** Various industries generate daily waste that is directly discharged into water bodies, causing harm. Industries such as paper manufacturing, textile dyeing, mills, steel production, chemical processing, and breweries produce substantial amounts of waste materials. These wastes contain harmful substances like strong acids (e.g., HCl), heavy metals, dyes, cyanides, DDT (Dichloro-Diphenyl-Trichloroethane), and suspended solids. When released into water bodies, they cause significant damage to aquatic life and local populations.
2. **Agricultural waste:** Chemicals and fertilizers used in agriculture also contribute to water pollution. Agricultural chemicals, synthetic fertilizers, and pesticides find their way into water bodies. These compounds contain high levels of nutrients like nitrates and phosphates, which promote the growth of algal blooms.
3. **Sewage waste:** Sewage waste, comprising household refuse, detergents, human and animal excrement, food remnants, and paper, forms the bulk of pollutants in water bodies. Its abundance overwhelms the capacity of aerobic microorganisms in water to break it down, leading to the demise of aquatic ecosystems.

Environmental Consequences of Water Pollution

1. Aquatic habitats are endangered by water contaminants, including various heavy metals, plastics, chemicals, and other harmful substances.
2. Excess nutrients in water bodies trigger algal blooms, resulting in oxygen depletion and creating hypoxic conditions that cause aquatic organism mortality.
3. Certain species sensitive to pollution struggle to survive in contaminated water, causing a reduction in aquatic biodiversity.
4. In some cases, pollution contributes to the disappearance of endemic or vulnerable species.
5. Water contamination from pollutants poses serious health risks to humans.

Biological Oxygen Demand (BOD)

The rise in water pollutants increases BOD as aerobic bacteria require more oxygen to degrade contaminants. However, pollutants hinder sunlight penetration and oxygen entry into water bodies, lowering oxygen levels and inhibiting aerobic microbial activity. The optimal BOD is 5; a higher value indicates insufficient oxygen.

1. Dissolved oxygen is measured in ppm; <0.2ppm indicates anoxic conditions.
2. Young fish require >6.0ppm for growth.
3. Adult fish need 3.0 to 5.0 ppm for spawning and growth.
4. <3.0 ppm is critically low for fish populations.

BOD levels under 5 are acceptable, while levels exceeding 5 are considered poor, and those above 100 are extremely poor. Excessive inorganic nutrients or algal blooms cause eutrophication, accelerating the natural aging of water bodies. Human activities are primarily responsible for eutrophication, which results in the death of numerous organisms, creates anoxic conditions, and reduces populations.

Impact of Water Pollution on Human Health

Water pollution leads to a scarcity of clean water for human consumption and causes various health issues and waterborne diseases. Research shows that individuals consuming fish from polluted waters face immediate and long-term health problems. This situation complicates and increases the cost of water treatment processes. The decline in fish populations affects marine life and impacts industries such as fishing, tourism, and recreation. Contaminated water also harms crops, reducing agricultural productivity. Pollutants diminish the aesthetic appeal of water bodies, decreasing their attractiveness to tourists. Elevated levels of chemicals, salts, and acids raise pH levels, posing risks to both aquatic life and human health.

I. Hudson River PCBs: A Case Study

From the late 1940s to 1977, General Electric (GE) released about 1.3 million pounds of polychlorinated biphenyls (PCBs) into the Hudson River from its facilities in Hudson Falls and Fort Edward, New York. Initially valued for their heat resistance in electrical equipment, PCBs were later discovered to be highly toxic and persistent pollutants. This pollution severely impacted the river's ecosystem, including its sediments, water, and aquatic organisms, creating health hazards for humans through the accumulation of PCBs in fish1. As a result, New York State shut down several fisheries and issued warnings about fish consumption in 1976.

The Hudson River was declared a Superfund site in 1984, prompting extensive EPA investigations. These studies determined that PCBs in the sediment posed a significant risk and that natural recovery was not sufficient1. The EPA approved a plan in 2002 to remove 2.65 million cubic yards of contaminated sediment, a project that was executed between 2009 and 20153. Despite these cleanup efforts, ongoing monitoring continues, and discussions about the effectiveness of the remediation persist, with some research indicating that additional dredging might be necessary4. The enduring effects of PCBs remain a concern, and the EPA works alongside state and local partners to facilitate the river's recovery.

II. The Cuyahoga River Fire (1969)

The Cuyahoga River in Cleveland, Ohio, suffered from extensive pollution due to years of industrial discharge and oil waste. On June 22, 1969, a spark from a passing train ignited the oil and debris in the river, resulting in a fire that lasted approximately 30 minutes and reached heights of up to five stories, causing minor damage to railroad bridges2. While river fires were not uncommon historically, the 1969 incident garnered nationwide attention, becoming a symbol of severe environmental degradation2. This event catalyzed the environmental movement, contributing to the establishment of the Environmental Protection Agency (EPA) and the enactment of the Clean Water Act in 1972, while emphasizing the necessity for more stringent environmental regulations and pollution control measures.

III. Tar Creek Superfund Site (Oklahoma)

The Tar Creek Superfund Site, located in northeastern Oklahoma, was a significant lead and zinc extraction region from the early 1900s through the 1960s. The mining activities resulted in substantial waste accumulations and contaminated water resources with toxic metals such as lead, zinc, cadmium, and arsenic. Following the cessation of mining operations, water infiltrated the abandoned mines, becoming acidic and extracting pollutants, which subsequently contaminated the surrounding land and water systems.

The environmental and health ramifications of the site have been profound, with lead poisoning in children being particularly concerning, leading to enduring health issues. It ranks among the most dangerous Superfund sites in the United States. Cleanup efforts involve extracting contaminated soil, covering waste mounds, and purifying water, but obstacles remain due to the extensive pollution and the vast area affected. These instances underscore the importance of appropriate industrial waste handling and environmental regulations to avert such catastrophes.

2. Air Pollution

The excessive use of natural resources by humans to satisfy their greed and desire for comfort has severely impacted the environment. This reckless exploitation has led to the extinction of numerous endangered, threatened, and vulnerable species. As a consequence, people are now suffering from life-threatening chronic diseases. The widespread use of polluting vehicles, including expensive cars and showy motorcycles, significantly contributes to air pollution. Furthermore, the uncontrolled release of industrial emissions into the atmosphere degrades air quality and contaminates the environment, causing harm to both local natural resources and the health of residents and workers in the area. Additional activities, such as smoking and the use of substances that increase emissions of SO_2, CO, and other harmful gases, also exacerbate air pollution. The primary cause of air pollution is gaseous emissions, which originate from various sources, including:

1. Industrial emissions
2. Automobiles
3. Domestic combustion
4. Agricultural activities
5. Burning fossil fuels.

Air pollution stems from both natural and anthropogenic causes. Emissions from factories, power plants, vehicles, and other industrial activities are major contributors. The air contains high levels of sulfur oxides, nitrogen oxides, and volatile organic compounds, making it toxic and difficult to breathe, resulting in various chronic illnesses. The two main categories of air pollution causes are:

I. Man-Made causes of pollution
II. Natural causes of air pollution

I. Man-Made Causes of Pollution

Population growth has led to an increased demand for vehicles, with many individuals opting for personal transportation over public transit. This surge in vehicle numbers results in higher emissions of pollutants such as carbon monoxide, nitrogen oxides, hydrocarbons, and particulate matter. It also leads to increased exhaust fumes from diesel engines. The combustion of oil, coal, and natural gas for energy production releases substantial amounts of carbon dioxide, sulfur dioxide, and other pollutants. Many power plants rely on these fossil fuels for energy generation. The growing population's need for food and shelter has led to increased land use, resulting in deforestation and the burning of forest wood. This process releases significant amounts of carbon dioxide and other pollutants, further contributing to air pollution. The use of pesticides, herbicides, and fertilizers in agriculture releases harmful chemicals into the air, posing health risks to workers who inhale these substances. Open burning of agricultural and domestic waste, along with emissions from landfills, releases large quantities of pollutants such as methane and carbon dioxide.

II. Natural Cause of Air Pollution

Volcanic activity can release substantial quantities of gases and ash into the atmosphere, resulting in short-term cooling effects due to reduced sunlight penetration and causing breathing difficulties for organisms. The discharged ash can also interfere with air travel and influence climate patterns. Forest fires produce considerable amounts of smoke and particulates, which may lead to respiratory issues, diminish visibility, and contribute to climate change by releasing carbon stored in trees and plants. Radon, a naturally occurring radioactive gas that emerges from the ground, particularly in uranium-rich areas, can accumulate in structures and is a primary cause of lung cancer after tobacco use. Oceans emit sea salt particles into the air through wave action and evaporation. These particles can affect cloud formation and weather patterns, and while generally harmless, they contribute to natural background levels of particulate matter. Geothermal vents and hot springs emit gases such as hydrogen sulfide, which can cause breathing problems and play a role in the natural sulfur cycle. Air pollution significantly affects the environment by reducing air quality, damaging ecosystems, and contributing to climate change. Moreover, air pollution leads to phenomena such as acid rain and global warming, further harming natural habitats and increasing stress on environmental systems.

Acid Rain

Elevated levels of nitric and sulfuric acids form precipitation that results in acid rain. This rain is not typical, having a pH of approximately 4.2-4.4, which is mildly acidic due to increased hydrogen ion concentrations. This pH is considerably lower than that of normal rain, which has a pH of 5.6. Acid rain results from high precipitation of sulfuric and nitric acids originating from both natural and anthropogenic sources. This type of rain is harmful, potentially causing various skin issues, irritation, and other negative effects. Acid rain forms when SO_2 and NO_x are emitted into the atmosphere. These compounds undergo photooxidation in the presence of sunlight and oxygen upon release. They then interact with water and other atmospheric components to produce sulfuric acid (H_2SO_4) and nitric acid (HNO_3). These acids are transported by wind and air currents, eventually falling as wet deposits (rain, fog, snow) or dry deposits (dust, gas, smoke). Acid rain has numerous detrimental effects, including the corrosion of structures and statues made from limestone and marble. It renders water bodies unsuitable for aquatic life, making many rivers and lakes unfit for use. Additionally, it depletes essential nutrients like calcium and magnesium from the soil while releasing aluminum, which can impede plant growth. Acid rain harms trees by depleting soil nutrients and causing direct harm to foliage and tree bark. When acid rain comes into contact with human skin, it can result in various dermatological issues, causing irritation to both the eyes and skin, and creating a sense of discomfort. Some methods to mitigate the effects of acid rain include wearing clothing that covers the arms, utilizing air purification devices, and supporting clean air initiatives. It is important to reduce the direct release of nitric oxide and sulfur oxide waste into the atmosphere by employing purifier chimneys and fume hoods.

Global Warming

The Earth's increasing temperatures, known as global warming, are caused by the emission of greenhouse gases such as carbon dioxide, methane, and nitrous oxide. These gases are released into the atmosphere through the combustion of fossil fuels, deforestation, and industrial emissions, all of which contribute to global warming. Greenhouse gases function by trapping heat within the Earth's atmosphere, preventing it from escaping into space, a process known as the greenhouse effect. These gases play a crucial role in maintaining the Earth's atmosphere and retaining heat, as many organisms require a stable temperature for survival. Without greenhouse gases, Earth's temperature could potentially drop below -18 degrees Celsius, making it inhospitable for numerous species. While greenhouse gases are a natural phenomenon that protects Earth's inhabitants, human activities that increase their release can lead to global warming. This phenomenon has numerous detrimental effects on the environment and ecosystem. Elevated temperatures cause glaciers to melt, contributing to rising sea levels and potential flooding. Higher temperatures can also result in extreme weather events such as hurricanes, droughts, and heavy rainfall. Additionally, increased temperatures can lead to health issues like heat-related illnesses and respiratory problems. Biodiversity is also at risk, as rising temperatures may cause the extinction of various endangered species. Furthermore, global warming can reduce agricultural productivity and increase healthcare costs, as many medical supplies and equipment require storage at very low temperatures, such as vaccines, organs, and samples. Transporting these items during emergencies will become more expensive. Efforts to combat global warming include reducing greenhouse gas emissions through the use of renewable energy sources, improving energy efficiency, and promoting reforestation. Adaptation strategies involve preparing for the impacts of global warming, such as constructing resilient infrastructure and developing crops that are resistant to drought. The future of global warming depends on the actions taken today. Reducing emissions and transitioning to sustainable practices are essential for mitigating its effects and ensuring a stable climate for future generations.

3. Land Pollution

Land degradation resulting from anthropogenic activities encompasses several key factors:

1. **Industrial Operations:** Manufacturing facilities generate waste products, including chemical compounds, metallic elements, and various contaminants that compromise soil integrity. Inadequate disposal of industrial byproducts, particularly toxic substances from production processes, exacerbates land pollution.
2. **Agricultural Practices:** The application of agrochemicals, including pesticides, herbicides, and fertilizers, contributes to soil deterioration and contamination.
3. **Forest Clearance:** The removal of woodland areas for agricultural expansion, urban development, or timber harvesting diminishes the land's natural regenerative capacity and promotes soil erosion.
4. **Urban Expansion:** Population growth necessitates increased land utilization, leading to rapid urban development. The construction of edifices, roadways, and infrastructure disrupts natural landscapes and fosters pollution. Construction sites produce particulate matter and debris, while urban areas often struggle with improper waste management.
5. **Radioactive Waste:** The mishandling of nuclear byproducts poses significant hazards to surrounding populations.
6. **Chemical Incidents:** Unintentional releases of hazardous substances, such as petroleum products, solvents, and other noxious materials, can contaminate both soil and water resources.
7. **Plastic Accumulation:** Improper disposal and buildup of plastic waste result in persistent land pollution. Discarded plastic items litter the environment and degrade into microplastics, further contaminating soil ecosystems.

These primary contributors to land pollution have far-reaching implications for ecosystems, human well-being, and biodiversity. Mitigating these extensive pollution sources is crucial for ensuring a sustainable future.

Case Study

Land degradation in Punjab

Punjab, an agrarian region in India, grapples with significant land degradation issues stemming from unsustainable farming methods, industrial effluents, extensive deforestation, and accelerated urbanization coupled with population growth. These factors have led to biodiversity loss, soil degradation due to excessive agrochemical use, and various health concerns. Implementing sustainable agricultural practices, such as transitioning to organic farming and reducing reliance on synthetic pesticides and fertilizers, is essential. Such measures would help prevent water contamination and associated health risks. Additionally, reforestation initiatives can restore natural habitats and enhance soil quality.

Land Fill Pollution in Mexico

Mexico City, one of the world's most densely populated urban centers, faces significant challenges in municipal waste management, resulting in landfill-related issues and environmental degradation. The city struggles with inadequate waste management practices, leading to improper disposal in landfills. Limited public awareness and insufficient recycling infrastructure result in the disposal of numerous recyclable materials in landfills. Consequently, substantial quantities of leachate are generated from decomposing organic waste and rainwater percolation. This leachate contains various harmful chemicals and heavy metals, contributing to soil erosion. Furthermore, the decomposition of organic waste in landfills produces methane, a potent greenhouse gas that exacerbates global warming. Methane emissions also present fire and explosion hazards, compounding the environmental and safety concerns associated with inadequate waste management practices.

Noise Pollution

An Environmental Concern pollution, commonly known Acoustic pollution, refers to excessive or disruptive sounds that interfere with normal environmental activities and cause disturbances. Human-generated activities are the primary contributors to this phenomenon. The tendency to disrupt others for personal enjoyment and self-interest has become a prevalent human behavior.

Origins of Noise Pollution

1. **Amplified Audio:** Urban development and leisure pursuits have led to an increase in avoidable sources of noise pollution, including late-night gatherings, public events, loudspeakers, amplified music, crowd noise,

announcements, sound systems, and audience-generated sounds.

2. **Domestic Sources:** Household appliances, such as kitchen equipment, laundry machines, televisions, and stereos, as well as renovation activities, contribute to residential noise pollution.
3. **Agricultural Operations:** Noise generated by fieldwork, pumping, and irrigation, particularly in large-scale farming, adds to acoustic pollution.
4. **Construction Activities:** Significant noise pollution stems from construction work, including the operation of jackhammers, bulldozers, cranes, and machinery used in road maintenance, construction, and demolition. These sounds are highly disruptive and produce vibrations that affect numerous individuals in proximity.
5. **Industrial Processes:** Manufacturing facilities, construction sites, and heavy machinery in industrial zones contribute substantially to noise pollution.
6. **Transportation-Related Noise:** Sounds emanating from road traffic, public transit, railway signals, track friction, and aircraft takeoffs and landings all contribute to acoustic pollution.

Impact of Noise Pollution

The impact of noise pollution extends beyond human discomfort. Many animal species rely on acoustic signals for communication, navigation, predator avoidance, and reproduction. Excessive noise impairs their ability to communicate effectively, leading to difficulties in habitat location, mate selection, and territory defense. Some avian species abandon their habitats due to excessive noise levels. Prolonged exposure to noise can induce stress in animals, resulting in physiological changes, compromised immune function, and heightened threat perception. Persistent acoustic pollution can trigger large-scale population migrations and contribute to the extinction of various biodiverse species. This can also lead to the overcrowding of animals or insects in quieter areas, intensifying competition for natural resources and challenging survival. Underwater noise pollution from maritime vessels, sonar equipment, and drilling operations can cause physical trauma, disorientation, and mating difficulties for aquatic organisms. In some instances, this leads to mortality among various aquatic species.

Case Study: Underwater Noise Pollution in the Mediterranean Sea

The Mediterranean Sea, a heavily trafficked maritime area, experiences significant noise pollution from shipping, recreational boating, and underwater construction activities. Research has shown that this acoustic disturbance can adversely affect the behavior and health of marine species, including fish and invertebrates. Studies have demonstrated that noise pollution can induce stress, hearing damage, and alterations in reproductive and feeding behaviors among marine life.

Case Study: Noise Pollution from Seismic Surveys in the Gulf of Mexico

Seismic surveys conducted for oil and gas exploration employ air guns to generate intense sound waves that penetrate the seabed. These loud noises can inflict physical harm on marine mammals, such as whales and dolphins, resulting in hearing loss, stress, and behavioral changes. Research has revealed that seismic surveys can disrupt the communication and navigation of these animals, impacting their ability to locate food sources and potential mates.

Public Health Concerns

The escalating prevalence of pollution and unsustainable human practices has precipitated a cascade of adverse effects on ecosystems, wildlife, and human populations. Anthropogenic activities that exploit natural resources have engendered numerous deleterious consequences for public health. This discourse will elucidate the manifold health challenges arising from environmental degradation.

Pollution-Induced Health Hazards

1. Atmospheric pollutants, including nitrogen oxides, sulfur dioxide, particulate matter, and volatile organic compounds such as benzene and formaldehyde, are associated with a spectrum of respiratory, cardiovascular, and oncological disorders in younger demographics.
2. The proliferation of waterborne pathogens due to contaminated water sources has led to outbreaks of cholera and dysentery. Moreover, the presence of heavy metals, agrochemicals, and industrial effluents in aquatic ecosystems has been linked to neurological impairments, carcinogenesis, and reproductive anomalies.
3. The bioaccumulation of toxic substances in crops has resulted in the inadvertent consumption of harmful chemicals. Furthermore, the infiltration of microplastics into the food chain and the indirect ingestion of heavy metals have been implicated in renal dysfunction.
4. Chronic exposure to noise pollution exceeding 85 decibels has been correlated with auditory impairment, hypertension, and cardiovascular morbidity. Additionally, nocturnal disturbances due to excessive noise

have been associated with sleep disorders and compromised overall well-being.

4. Solid Waste Pollution

Improper solid waste management has emerged as a significant environmental concern, with far-reaching implications for terrestrial, atmospheric, and aquatic ecosystems.

Primary Sources of Solid Waste

1. **Industrial sector:** Manufacturing processes generate substantial quantities of non-biodegradable and non-recyclable waste, including chemical residues and discarded packaging materials.
2. **Residential areas:** Household waste encompasses a diverse array of materials, including organic matter, paper products, plastics, glassware, and metallic objects.
3. **Construction and demolition:** This category comprises discarded building materials such as concrete, bricks, lumber, and metal scraps.
4. **Agricultural activities:** Crop residues, pesticide remnants, surplus fertilizers, and animal excreta constitute the primary forms of agricultural waste.
5. **Electronic waste:** Obsolete electronic devices and their components, including computers, mobile phones, and televisions, contribute significantly to this waste stream.
6. **Healthcare facilities:** Medical waste encompasses used equipment, dressings, surgical instruments, and pharmaceutical byproducts.
7. **Mining operations:** Extractive industries generate substantial quantities of waste, including tailings and slag from mineral and metal processing.

Implications of Solid Waste Contamination

The infiltration of deleterious substances from waste into the earth can detrimentally affect soil productivity and plant growth, leading to soil degradation. Effluent from waste disposal facilities has the potential to contaminate both subterranean and surface water sources. The combustion of waste releases toxic gases and particulates into the atmosphere. Improperly managed waste can attract disease vectors, such as rodents and insects, facilitating pathogen transmission. Exposure to hazardous waste materials may result in health complications, including respiratory distress, cutaneous irritation, and potential carcinogenic effects. Waste containing sharp objects and toxic substances poses a risk of physical injury. Substantial financial resources are necessary for the management and remediation of contaminated areas. Regions with significant waste pollution often experience a decline in real estate values. Addressing solid waste contamination requires a comprehensive strategy involving individuals, communities, industries, and governmental entities. Through the implementation of sustainable practices and efficient waste management techniques, it is possible to mitigate the adverse effects of solid waste pollution on the environment and human health.

Disaster Management

Disaster management encompasses a coordinated approach to addressing emergencies and catastrophic events, aiming to minimize their impact on human life, property, and the environment. It involves a series of measures and strategies for preparation, response, recovery, and mitigation of disaster effects. The concept of disaster management has evolved over time without a single originator. However, the modern approach to disaster management began to take shape with the emergence of civil defense during the 20th century, particularly influenced by World War II experiences, which highlighted the need for organized responses to large-scale emergencies. The concept has since expanded to include disaster risk reduction (DRR), focusing on prevention and mitigation, and disaster management, involving response and recovery efforts. International organizations such as the United Nations Office for Disaster Risk Reduction (UNDRR) and the International Federation of Red Cross and Red Crescent Societies (IFRC) have played crucial roles in shaping and promoting global disaster management practices. National Agencies: Governments often establish dedicated disaster management agencies (e.g., FEMA in the United States, NDMA in India) responsible for coordinating and implementing disaster response and recovery efforts.

Effective disaster management necessitates collaboration among governments, organizations, communities, and individuals. Through preparation and mitigation efforts, it is possible to safeguard lives, protect property, and facilitate rapid recovery for affected communities.

Types of Disasters Management

1. **Natural Disaster** - This category encompasses events that occur naturally, originating from environmental phenomena. It includes:

    i. **Geological disasters**, such as seismic activity, mass movements, tsunamis, and volcanic eruptions.

    E.g., Eruption of Mount Vesuvius in 79 AD, Italy.

ii. **Hydrologic disasters**, including phenomena such as droughts and flash floods.

E.g., Dust Bowl in U.S (1930)

iii. **Meteorological Disaster** - this category comprises hurricanes, tornadoes, blizzards, and heat waves.

E.g., Hurricane Katrina in US.

iv. **Biological disasters -** Pandemic situations

Coronavirus pandemic globally (2019).

2. **Human-Induced Disasters** - Calamities resulting from anthropogenic activities. These encompass:

I. **Incidents** - Chemical releases, detonations, radioactive leakages, transport-related accidents.

II. **Ecological contamination** - Discharge of hazardous substances, petroleum in oceanic and littoral regions, emission of noxious gases.

III. **Sociopolitical Catastrophes** - Warfare leading to widespread devastation, familial disintegration, and terrorist actions.

Importance of Disaster Management

1. Safeguarding individuals and societies by diminishing disaster impacts.
2. Curtailing damage to assets and infrastructure, thus reducing fiscal losses.
3. Conserving natural environments and ecosystems vulnerable to disasters.
4. Fostering adaptable communities capable of swift post-disaster recovery and enhanced preparedness.

Very Short Answer Type Questions

1. Explain water pollution and its cause.
2. How industrial effluents contributes towards polluting the environment.
3. What do you understand by the term "Biological Oxygen Demand"
4. What is considered as good BOD?
5. Define eutrophication with example.
6. What are the natural causes of air pollution?
7. Describe the phenomenon of acid rain.
8. Discuss any three sources of noise pollution.
9. Describe any two consequences of environmental pollution.

Short Answer Type Questions

1. Discuss the impact of water pollution on environment.
2. How BOD leads to the problem of eutrophication.
3. Discuss the case study of Cuyahoga River fire in brief.
4. What are the main causes of air pollution.
5. What is the impact of elevated level of green house effect.
6. Discuss the various sources of noise pollution.
7. Discuss any case study of noise pollution.
8. What are the consequences of solid waste contamination.
9. Discuss government employed disaster management agencies.
10. How man-made causes are responsible for disaster management.

Long Answer Type Questions

1. Discuss disaster management in detail.
2. Describe the harmful effects of global warming and how it is caused?
3. What are the health problems caused due to acid rain and how they can be prevented?
4. Describe a case study each on water pollution and land pollution.
5. Describe the causes of land pollution.
6. What are the harmful impact of noise pollution on environment. Discuss in detail.
7. How pollution can be detrimental for human health.
8. Describe solid waste management in detail.
9. What is the Disaster management and why it is important.

Multiple Choice Questions

1. Cuyahoga river fire occured

   a. 1980 **b. 1969**

   c. 1950 d. 1996

2. Burning of fossil fuels cause which type of pollution

   a. Land pollution **b. Air pollution**

   c. Solid waste pollution d. Water pollution

3. Acid rain occurs due to precipitate of

   a. Nitrogen oxide and sulphur oxide. **c. Nitric acid and sulphur oxide**

   c. Nitric acid and sulphuric acid d. None of the above

4. Global warming is caused due to

   a. High elevated temperature b. Emission of greenhouse gases

   c. Due to low temp d. None of the above

5. Air pollution is caused due to
    a. Listening loud music
    b. Household chores
    c. Industrial effluents
    d. Industrial emissions
6. Dysentery is caused due to
    a. Air pollution
    b. Solid waste disposal
    c. Water pollution
    d. Land fills
7. Noise pollution can cause
    a. Health problems
    b. Hurdle in mating
    c. Both
    d. None of the above
8. Hearing voice above 85 decibel will lead to
    a. Hearing loss
    b. Mental stress
    c. Increased blood pressure
    d. Both ii and iii.
9. What is the ph of normal rain
    a. 7
    b. 5.6
    c. 5.9
    d. 4.2
10. Solid waste can be managed through
    a. Proper waste management
    b. Recycling paper waste
    c. By filling undegradable waste into landfills
    d. All of the above

9

Pollution and Its Impact on Environment

Pollution, a significant concern in the contemporary world due to its detrimental effect on the quality of life, has been experiencing a substantial increase as a result of the considerable expansion and intensification of industrialization and urbanization. Pollution is now widely acknowledged as a major source of economic inefficiency and severe health problems. It is estimated that in any given period, more than 100 million tons of pollutants of various classifications are released into the atmosphere. The escalating levels of pollution are considered to be the direct consequence of advancements in technology and industrialization, which are degrading the environment at an alarming rate. The impact of pollution is such that now almost everyone accepts the inevitability of anthropogenic pollutants in the environment; human life and economic development now evolve around this polluted environment. Numerous individuals succumb to air pollution annually, and hundreds of millions more suffer from stunted growth, non-fatal bronchitis, and other industrial-induced problems. The majority of the world's water is contaminated by organic waste and heavy metals. Some countries have water that is acutely polluted, and many have medium to high levels of pollution. It is posited that over 10,000 people die daily as a result of this pollution. The economics of the environment and economy are themselves distorted by polluting technologies. Ambitious control programs run concurrently with increasing temperatures, acid rain, strip deserts, and depleted fisheries and forests. Countries are compelled to increase public spending in areas of pollution, thereby reducing the availability of funds for other, more pressing needs.

Types of Pollution

There are various types of pollution, including agricultural, industrial, domestic, chemical, thermal, light, noise, taste, visual, and radioactive pollution. The organisms that cause pollution can be classified into two main categories: point and non-point sources. The point sources are outlets where pollution occurs as a result of a single or few events. They are easily identifiable and measurable. They include power and industrial plants, sewage treatment plants, and factories. The non-point pollution sources are not easily quantifiable or definable. They occur as a result of runoff of chemicals such as pesticides and fertilizers, chlorine from water cleansers, and other concentrated discharges.

There are also various sources of pollution. These include motor vehicle exhaust, aircraft exhaust, chimney exhaust from power and industrial plants, household and industrial chemicals, and radioactive sources. Pollution renders the environment unsuitable for fauna such as birds and fish, as well as flora and humans. The environmental polluting substances can be classified into two main categories, namely physical and biological pollutants. The physical pollutants include sunlight, heat, noise, and chemicals. The biological pollution stems from the growing amount of waste produced by people. In order to mitigate pollution, all these categories of pollutants need to be addressed in a comprehensive manner. Environmental pollution is a global problem that necessitates global solutions. Environmental pollution is a serious worldwide issue. The pollutants form as a result of a chemical reaction. Some types of pollution we have already discussed in chapter 7; in this chapter we will examine solid pollution in more detail.

Solid Waste Pollution

An increasing concern in contemporary society is the environmental impact of improper solid waste disposal. Historically, landfill techniques have been employed for waste management; however, these methods result in air and water pollution, depletion of valuable resources, and are unsustainable in the medium to long term. Consequently, incineration of solid waste for energy recovery has emerged as an alternative approach. Nevertheless, the by-products of the incineration process comprise a mixture of residues, including fly ash, dense ash, clinkers, and flue gas, which serve as a vector for heavy metal emissions. These products require processing to mitigate their potential detrimental effects. A patented process for eliminating waste residues from the environment, which integrates thermal treatment of waste residues with concrete manufacturing, has been developed. Solid waste also generates malodorous emissions due to rapid fermentation and becomes problematic when dispersed by wind or improper handling.

Sources of Solid Waste Pollution

1. Industrial activities
2. Agricultural Practices
3. Municipal Waste

1. Industrial Activities

Industrial processes generate substantial quantities of waste. For instance, pulp and paper mills, oil refineries, sugar factories, and glass industries produce significant amounts of solid waste. Many industries engage in improper disposal practices, such as dumping waste on land or in water bodies. This can lead to contamination of soil and water sources with harmful substances, including heavy metals, toxic chemicals, and non-biodegradable materials. Industries frequently utilize chemicals in their processes, which can become hazardous waste. These chemicals may include solvents, acids, alkalis, and other toxic substances. Industrial activities also produce emissions and byproducts that contribute to solid waste pollution. For example, the combustion of fossil fuels releases particulate matter and other pollutants that can settle on the ground and contribute to solid waste accumulation. Many industries lack effective recycling or reuse programs for their waste materials, resulting in an increase in the volume of solid waste requiring disposal and exacerbating the pollution problem. Moreover, these practices alter soil layer strength, reduce soil fertility, and diminish biological activity. When waste is deposited in soil, it can contaminate the soil and render it unsuitable for agriculture. Additionally, waste that enters water bodies can render the water unsafe for drinking and other domestic uses.

2. Agricultural Practices

Crop residues, such as straw, stalks, and husks, are often left in fields post-harvest. If not managed appropriately, these residues can become waste. The practice of burning these residues is common but leads to air pollution, while leaving them to decompose can attract pests and diseases. The utilization of plastic products, including mulch films, greenhouse coverings, and packaging, is widespread in agriculture. These plastics can become waste when they are no longer functional. Improper disposal can result in soil and water pollution. Animal farming produces significant quantities of waste, including manure, bedding material, and carcasses. Inadequate management of livestock waste can lead to water pollution, offensive odors, and the proliferation of pathogens.

Agricultural processing generates organic waste, including fruit and vegetable remnants, which can attract pests and contribute to pollution if mismanaged. The excessive use of packaging materials, particularly non-biodegradable ones, exacerbates solid waste pollution. Furthermore, improper agricultural practices may result in soil erosion, leading to sedimentation in waterways and subsequent impacts on aquatic ecosystems and water quality.

3. Municipal Waste

Municipal waste encompasses a diverse range of materials from commercial, household, and institutional sources, including food scraps, plastics, garden waste, office paper, medical waste, and packaging materials. Food waste, prone to decomposition, originates from animal and vegetable products during various stages of food handling and consumption. Non-decomposable municipal waste is primarily disposed of in landfills or dumpsites, which can lead to soil and water contamination through leachate and methane production. Littering and illegal dumping further contribute to environmental degradation, harming wildlife and natural habitats. Incineration, while reducing waste volume, releases harmful pollutants such as dioxins and furans into the atmosphere. A significant portion of municipal waste consists of recyclable materials, which, when improperly managed, contribute to solid waste pollution and resource depletion. Organic waste, including food scraps and garden waste, produces methane when decomposing anaerobically in landfills. The growing presence of electronic waste in municipal waste streams poses additional environmental risks due to its hazardous components.

Ways to Mitigate Solid Waste Pollution

Strategies to mitigate solid waste pollution include:

1. Implementation of proper waste management practices in industries
2. Public awareness campaigns and enhanced environmental regulation enforcement
3. Promotion of agricultural waste recycling and reuse
4. Safe disposal of chemical containers and hazardous materials
5. Utilization of biodegradable or reusable materials
6. Establishment of effective waste management systems in agricultural settings
7. Adoption of sustainable farming practices
8. Encouragement of waste reduction at the source
9. Expansion and improvement of recycling programs
10. Promotion of organic waste composting to reduce methane emissions

Case study to reduce solid waste pollution

Pune, India

Pune, an Indian city experiencing rapid growth, encountered substantial difficulties in managing solid waste due to its expanding population and urban development. The city produced considerable amounts of municipal solid waste, including plastics, which created environmental and health hazards. To address this issue, Pune established innovative collaborations with informal waste collectors to reclaim plastic waste. This inclusive circular economy strategy connected informal and formal waste processes. The city embraced a hybrid system that incorporated informal workers alongside a formal waste management structure, which was regarded as a success in the field.

Pune initiated extensive public education campaigns to inform residents about waste separation, recycling, and waste reduction importance. Changes at the national level in municipal solid waste management policies supported Pune's efforts to revolutionize its waste handling system. These initiatives resulted in significant environmental improvements, such as decreased plastic waste in landfills and enhanced recycling rates. The integration of informal workers into waste management provided them with steady employment and enhanced their quality of life. Citizen participation in waste management activities grew, resulting in a cleaner and healthier urban environment.

This case study illustrates that involving informal workers and the community in waste management efforts can yield positive outcomes. It emphasizes the importance of public education about waste management practices for achieving long-term sustainability. Additionally, it highlights the necessity of robust policy frameworks and national support in transforming waste management systems. The study demonstrates how innovative approaches and community engagement can effectively reduce solid waste pollution and foster a more sustainable environment.

Bengaluru, India

To address the increasing solid waste pollution, Bengaluru has launched innovative green initiatives. Plogging, which combines jogging with litter collection, is organized by the Indian Ploggers Army and various volunteers in parks, lakes, and public spaces, collecting significant waste such as plastics, bottles, and cups. This activity promotes cleanliness, fitness, and waste management awareness. Compost santhe markets sell compost from organic waste, encouraging household composting and reducing landfill waste while educating the public on composting benefits. Zero-waste events and weddings minimize waste by using reusable or biodegradable items and proper waste segregation. These practices reduce environmental impact and set sustainable examples.

Recycling programs in Bengaluru include recycling centers, promoting recycled products, and encouraging sustainable business practices, reducing landfill waste and conserving resources. Public awareness campaigns using media such as radio, TV, and social platforms educate residents on waste segregation, recycling, and reduction, changing public behavior and encouraging participation. Policy changes, including plastic regulations and recycling incentives, support Bengaluru's waste management, creating a sustainable framework and encouraging compliance.

Community engagement involves residents, volunteers, and organizations in clean-up drives, awareness campaigns, and recycling programs, fostering responsibility and ownership in waste management. These combined efforts have reduced plastic waste, increased recycling rates, and heightened public awareness, providing economic opportunities for informal waste workers and promoting a cleaner environment. Educating the public and engaging informal waste workers are crucial for sustainable outcomes, supported by strong policy frameworks and national backing.

Very Short Answer Type Questions

1. Define pollution
2. What are the different types of pollution you observe in your surroundings?
3. Discuss solid waste pollution.
4. Name the different sources of pollution.
5. Tell at least two ways to regulate pollution.

Short Answer Type Questions

1. Define solid waste pollution and its cause?
2. Discuss how industrial activities contribute to solid waste pollution.
3. How can we reduce the solid waste pollution?
4. Municipal waste is the major contributor of solid waste. Comment.
5. Discuss a case study to reduce solid pollution.

Long Answer Type Questions

1. Discuss different types of pollution.
2. What are the major sources of solid waste pollution? Discuss in detail.
3. What is the environmental impact of these polluting bodies? Discuss in detail.
4. Discuss the ways to mitigate the effect of solid pollution.
5. What measures were taken in Pune to prevent pollution?
6. Discuss two case studies to reduce environmental pollution.

Multiple Choice Questions

1. Pollution is classified into -
   a. Point sources
   b. Non- point sources
   c. Both
   d. None
2. Which comes under the category of Physical pollutants?
   a. Air
   b. Water
   c. Land
   d. Sunlight
3. What are the contributors of biological pollution.......
   a.
   b.
   c. Man-made activities
   d.
4. What are the sources of Solid waste pollution
   a. Industrial
   b. Agricultural
   c. All of them
   d. None of them
5. Ways to mitigate solid waste pollution are
   a. Use plastic bags
   b. Use Non-renewable substances
   c. Implementation of proper waste management practices
   d. Use of recyclable products

10

Climate Change Ecology

Introduction to Climate Change Ecology

Climate change ecology is the scientific study of how shifts in the Earth's climate system influence living organisms, their environments, and the complex interactions that bind them together. As the global climate changes—driven primarily by human activities such as burning fossil fuels, deforestation, and large-scale agriculture—the natural world is experiencing profound and accelerating transformations. Understanding these changes is essential for predicting the future of ecosystems, conserving biodiversity, and ensuring the continued provision of the ecosystem services upon which human societies depend.

The Climate System and Its Influence on Life

The Earth's climate is shaped by the balance of incoming and outgoing energy, regulated by atmospheric gases, ocean currents, and the planet's surface characteristics. The greenhouse effect, a natural process in which certain gases trap heat in the atmosphere, keeps the planet warm enough to support life. However, human activities have significantly increased the concentrations of greenhouse gases, especially carbon dioxide, methane, and nitrous oxide. This has enhanced the greenhouse effect, leading to a rise in average global temperatures and more extreme weather patterns.

Climate is a fundamental driver of ecological processes. It influences where species can live, how populations grow or decline, and how ecosystems function. Temperature, precipitation, and seasonal cycles determine the distribution of plants and animals, the timing of biological events like flowering and migration, and the productivity of ecosystems. As climate shifts, so too do the boundaries and characteristics of ecosystems, sometimes in abrupt and unpredictable ways.

Observed and Projected Ecological Changes

The ecological impacts of climate change are already evident across the globe. Rising temperatures have caused many species to shift their ranges toward the poles or to higher elevations in search of suitable conditions. For example, numerous bird and insect species now breed earlier in the year, and plants are flowering sooner than in the past. These changes can disrupt the delicate timing of ecological relationships, such as those between pollinators and the plants they service or between predators and their prey.

In aquatic environments, warming waters and ocean acidification—caused by the absorption of excess atmospheric carbon dioxide—are stressing coral reefs, altering fish distributions, and threatening the productivity of fisheries. Melting glaciers and shrinking sea ice are transforming polar ecosystems, endangering species that depend on ice-covered habitats.

Extreme weather events, such as heatwaves, droughts, floods, and wildfires, are becoming more frequent and intense. These disturbances can cause sudden and severe damage to ecosystems, leading to the loss of habitats, declines in populations, and even local extinctions. The expansion of deserts and the retreat of forests are visible signs of these changes.

Ecosystem Services and Human Well-being

Ecosystems provide essential services that support human life, including clean air and water, food, materials, and climate regulation. Climate change threatens these services by altering the composition, structure, and functioning of ecosystems. For instance, forests play a crucial role in absorbing carbon dioxide and regulating the global carbon cycle, but increased wildfires, pests, and diseases can reduce their capacity to act as carbon sinks.

Agricultural systems are also vulnerable, as changing temperatures and precipitation patterns affect crop yields and the availability of water. Coastal ecosystems, such as wetlands and mangroves, buffer communities from storms and flooding, but rising sea levels and increased storm intensity are eroding these natural defenses.

Interactions with Other Environmental Stressors

Climate change does not act alone; it interacts with other environmental pressures, such as habitat destruction, pollution, and the spread of invasive species. These combined stresses can push ecosystems beyond their limits, leading to rapid and sometimes irreversible changes. For example, forests already weakened by drought may be more susceptible to insect outbreaks or fire, while coral reefs facing both warming and pollution may be less able to recover from bleaching events.

Adaptation, Mitigation, and the Future of Ecology

Addressing the ecological impacts of climate change requires both mitigation—reducing greenhouse gas emissions to limit future warming—and adaptation—helping ecosystems and human communities adjust to new conditions. Conservation efforts, such as protecting and restoring habitats, managing invasive species, and maintaining ecological connectivity, can enhance the resilience of ecosystems.

Nature-based solutions, like reforestation and the restoration of wetlands, not only help absorb carbon but also provide co-benefits for biodiversity and human well-being. However, the success of these strategies depends on a deep understanding of ecological processes and the ability to anticipate and respond to ongoing changes.

Factors Affecting Climate Change

Climate change is shaped by a combination of natural processes and human activities that alter the balance of energy in Earth's atmosphere. While the climate has always varied due to natural causes, the rapid changes observed in recent decades are primarily the result of human actions. Understanding the factors that drive climate change is essential for developing effective strategies to mitigate its impacts.

A. Natural Factors

1. **Changes in the Earth's Orbit and Rotation:** Earth's position relative to the sun changes over thousands of years due to variations in its orbit and tilt. These cycles, known as Milankovitch cycles, influence the amount and distribution of solar energy the planet receives, leading to long-term climate patterns such as ice ages and interglacial periods.
2. **Variations in Solar Activity:** The sun's energy output fluctuates over time. Although these changes can affect the intensity of sunlight reaching Earth, recent measurements show that solar variations have played only a minor role in the warming observed over the past several decades.
3. **Volcanic Eruptions:** Volcanoes emit gases and particles into the atmosphere, some of which, like sulfur dioxide, can reflect sunlight and temporarily cool the planet. However, these effects are short-lived and do not explain the persistent warming trend.
4. **Natural Greenhouse Gases:** Water vapor, carbon dioxide, methane, and other greenhouse gases occur naturally and help regulate Earth's temperature. Before the industrial era, natural sources and sinks of these gases maintained a relatively stable climate.

5. **Ocean Circulation Patterns:** Large-scale movements of ocean water, such as El Niño and La Niña events, redistribute heat around the globe and cause periodic climate fluctuations. These patterns can influence global temperatures and weather, but they do not account for the long-term warming trend.

B. Human-Induced Factors

1. **Burning of Fossil Fuels:** The primary driver of recent climate change is the burning of coal, oil, and natural gas for energy and transportation. This releases large amounts of carbon dioxide and other greenhouse gases into the atmosphere, trapping heat and raising global temperatures.
2. **Deforestation and Land Use Changes:** Cutting down forests for agriculture, urban development, or logging reduces the planet's capacity to absorb carbon dioxide. When trees are removed and not replaced, stored carbon is released back into the atmosphere, further contributing to warming.
3. **Industrial Processes:** Manufacturing goods often involves chemical reactions that emit greenhouse gases. For example, producing cement releases carbon dioxide, and certain industrial activities emit potent gases like hydrofluorocarbons, which have a much greater warming effect than carbon dioxide.
4. **Agriculture and Livestock:** Agricultural practices, especially those involving livestock, release methane and nitrous oxide, both powerful greenhouse gases. Methane is produced during the digestion process in ruminant animals and from rice paddies, while nitrous oxide comes from fertilized soils.
5. **Transportation:** Most vehicles use fossil fuels, emitting carbon dioxide and other pollutants. The production and maintenance of transportation infrastructure, such as concrete for roads and bridges, also contribute to greenhouse gas emissions.
6. **Aerosols and Air Pollution:** Some pollutants, like sulfate aerosols, can reflect sunlight and have a temporary cooling effect. However, as air quality regulations have reduced these emissions, their masking effect on warming has diminished, allowing the full impact of greenhouse gases to be felt.
7. **Black Carbon (Soot):** Soot from burning fossil fuels and biomass absorbs sunlight and heats the atmosphere. When deposited on snow and ice, it accelerates melting by reducing the surface's ability to reflect sunlight.

Feedback Mechanisms and Amplifying Effects

Climate change is further complicated by feedback loops that can either amplify or dampen its effects:

1. **Melting Ice and Reduced Albedo**: As ice and snow melt, darker surfaces are exposed, absorbing more heat and accelerating further melting.
2. **Permafrost Thaw**: Warming temperatures cause permafrost to thaw, releasing methane and carbon dioxide that have been trapped in frozen soils, which in turn accelerates warming.
3. **Ocean Acidification**: Increased carbon dioxide is absorbed by the oceans, lowering pH levels and threatening marine ecosystems.

Patterns of Weather and Temperature

Weather and temperature patterns are fundamental aspects of Earth's climate system, shaping the environment, ecosystems, and human activities. Over recent decades, these patterns have undergone significant changes, driven largely by global warming and shifts in atmospheric dynamics. Understanding these patterns is essential for grasping both the immediate effects of weather and the long-term trends of climate change.

Weather and Temperature

Weather refers to the short-term conditions of the atmosphere at a specific place and time. It includes variables such as temperature, humidity, precipitation, wind, and atmospheric pressure. Weather can change rapidly, varying from day to day or even hour to hour.

Temperature is a measure of the warmth or coldness of the atmosphere as determined by a thermometer. It is one of the most direct and widely observed indicators of weather and climate. While weather describes short-term atmospheric conditions, **climate** refers to the average weather patterns over longer periods, typically decades or centuries.

Historical Patterns and Recent Trends

Historically, Earth's temperature and weather patterns have fluctuated due to natural factors such as volcanic eruptions, solar cycles, and changes in ocean currents. However, since the late 19th century, the average global temperature has risen by more than 1°C (about 2°F), with two-thirds of this warming occurring since the mid-1970s. This increase is closely linked to the rise in greenhouse gas concentrations from human activities.

The past decade has been the warmest on record, with each successive year frequently surpassing previous temperature records. The years 2023 and 2024, for example, have stood out as the hottest years since reliable thermometer measurements began, with global mean temperatures exceeding pre-industrial averages by more than 1.6°C. This warming trend has not been uniform: land areas are warming about twice as fast as oceans, and the Arctic is experiencing even more rapid temperature increases.

Variability in Weather Patterns

Rising average temperatures have been accompanied by significant changes in weather patterns, including:

1. **Increased Frequency of Heatwaves:** Extended periods of abnormally high temperatures are becoming more common, affecting human health, agriculture, and natural ecosystems.
2. **Changes in Precipitation:** Some regions are experiencing more intense rainfall and flooding, while others are facing prolonged droughts. The distribution and intensity of precipitation are shifting, often leading to more extreme weather events.
3. **Shifts in Storm Patterns:** Warmer ocean waters and atmospheric changes are influencing the frequency and intensity of storms, including hurricanes and cyclones.
4. **Altered Seasonal Cycles:** The timing of seasons is changing, with earlier springs, later autumns, and shifts in the onset and duration of seasonal weather patterns.

Regional Differences

Temperature and weather changes are not distributed evenly across the globe. Some regions, such as the northern United States, western North America, and Alaska, have experienced more pronounced warming. The Arctic is warming at more than twice the global average, leading to rapid ice melt and ecosystem changes. Conversely, some areas, like parts of the southeastern United States, have seen less warming or even periods of cooling, though these trends are generally less significant in the context of global averages.

Ocean and Land Temperature Patterns

The warming of the planet is evident both on land and in the oceans, but the patterns differ:

1. **Land:** Land surfaces heat up and cool down more quickly than oceans. As a result, land areas have experienced a sharper rise in average temperatures. This has led to more frequent and intense heatwaves, especially in urban areas where the "urban heat island" effect amplifies warming.
2. **Oceans:** Oceans absorb much of the excess heat from the atmosphere, moderating temperature increases but also leading to thermal expansion and sea level rise. Ocean surface temperatures have increased steadily, contributing to coral bleaching and changes in marine ecosystems.

Drivers of Recent Changes

The main driver of recent changes in weather and temperature patterns is the accumulation of greenhouse gases in the atmosphere, primarily from burning fossil fuels, deforestation, and industrial activities. These gases trap heat, leading to a gradual but persistent increase in global temperatures. Short-term fluctuations, such as those caused by El Niño and La Niña events, can temporarily amplify or dampen global temperature trends, but the long-term pattern remains one of steady warming.

Other factors influencing weather and temperature include changes in cloud cover, reductions in air pollution (which previously had a cooling effect), and shifts in atmospheric circulation. For example, reductions in sulfur aerosol pollution have allowed more sunlight to reach the surface, contributing to recent warming spikes.

Impact on Biodiversity and Ecosystem

Climate change is profoundly reshaping the natural world, with far-reaching consequences for biodiversity and the functioning of ecosystems. As global temperatures rise and weather patterns become more unpredictable, the delicate balance that sustains life on Earth is increasingly under threat. This impact is visible across terrestrial, freshwater, and marine environments, affecting species survival, ecosystem services, and the resilience of natural systems.

Shifts in Species Distribution and Extinction Risks

One of the most immediate effects of climate change on biodiversity is the forced movement of species. As temperatures climb, many animals and plants are migrating to higher elevations or toward the poles in search of suitable habitats. This shift disrupts existing ecological communities and can lead to the breakdown of long-established relationships between species. For example, pollinators may no longer coincide with the flowering times of plants, and predators may lose access to their traditional prey.

The risk of extinction increases with every degree of warming. Species that are unable to move quickly enough or adapt to new conditions face heightened danger. This is especially true for those with limited ranges, such as island species, or those already threatened by habitat loss and other human pressures. Recent decades have seen dramatic declines in wildlife populations, with some estimates suggesting that up to 20-30% of species could be at risk of extinction if temperatures rise by 1.5-2.5°C above pre-industrial levels.

Habitat Loss and Ecosystem Degradation

Climate change accelerates habitat loss through a variety of mechanisms. On land, rising temperatures, prolonged droughts, and more frequent wildfires can devastate forests, grasslands, and wetlands. The 2019-2020 Australian bushfires, intensified by climate change, burned vast areas of forest and destroyed habitats for countless species, increasing the number of threatened species in the region.

In marine environments, warming waters and ocean acidification are causing widespread damage. Coral reefs, among the most biodiverse ecosystems on the planet, are particularly vulnerable. Higher sea temperatures trigger mass coral bleaching events, while increased acidity makes it harder for corals and shellfish to build their skeletons. The loss of coral reefs not only threatens marine biodiversity but also undermines fisheries and coastal protection.

Sea level rise poses another threat, inundating coastal habitats such as mangroves and salt marshes. These ecosystems are crucial nurseries for many marine species and act as natural buffers against storms and erosion. As they disappear, both biodiversity and human communities become more vulnerable.

Altered Ecosystem Processes and Services

Ecosystems provide essential services that support life, including carbon sequestration, water filtration, soil formation, and pollination. Climate change disrupts these processes in multiple ways:

1. **Carbon Storage:** Forests, wetlands, and oceans act as major carbon sinks, absorbing and storing greenhouse gases. However, as these ecosystems degrade, their ability to sequester carbon diminishes, potentially accelerating climate change in a feedback loop.
2. **Water Regulation:** Changes in precipitation patterns and increased evaporation can reduce water availability, affecting both natural ecosystems and human agriculture.

3. **Nutrient Cycling:** Shifts in temperature and moisture alter the activity of soil microbes, impacting the decomposition of organic matter and the recycling of nutrients essential for plant growth.

The loss or decline of these services undermines food security, clean water supplies, and climate regulation, with direct consequences for human well-being.

Disruption of Biological Cycles and Interactions

Climate change is altering the timing of biological events, such as migration, breeding, and flowering. These phenological shifts can cause mismatches between species that depend on one another. For instance, if insects emerge earlier in the spring due to warmer temperatures but birds do not adjust their migration schedules, chicks may hatch when food is scarce.

Such disruptions ripple through ecosystems, reducing reproductive success and survival rates. Over time, these changes can lead to population declines and further loss of biodiversity.

Increased Frequency of Extreme Events

The frequency and intensity of extreme weather events—such as droughts, floods, storms, and heatwaves—are increasing. These events can devastate ecosystems, causing direct mortality, destroying habitats, and altering the structure and function of ecological communities. Recovery from such disturbances is often slow, and repeated events can push ecosystems beyond their capacity to recover, leading to long-term degradation.

Invasive Species and Disease

As climate zones shift, new opportunities arise for invasive species to establish themselves in previously inhospitable areas. These invaders can outcompete native species, disrupt ecosystem processes, and introduce new diseases. Warmer temperatures and altered precipitation patterns can also expand the range of pests and pathogens, further threatening biodiversity.

Human Land Use and Synergistic Effects

While climate change is a major driver of biodiversity loss, it often acts in concert with other human activities such as deforestation, agriculture, and urbanization. Land conversion for food production remains the leading cause of habitat loss, fragmenting ecosystems and reducing their ability to support diverse species. The combined effects of climate change and land use intensify the pressures on biodiversity, making it harder for ecosystems to adapt and recover.

The Role of Ecosystems in Climate Mitigation

Despite being under threat, healthy ecosystems are vital allies in the fight against climate change. Forests, peatlands, mangroves, and seagrasses absorb and store vast amounts of carbon, helping to regulate the global climate. Protecting and restoring these natural carbon sinks is essential not only for conserving biodiversity but also for limiting future warming.

Conclusion

The impact of climate change on biodiversity and ecosystems is profound and accelerating. Species are being pushed beyond their limits, habitats are being lost or transformed, and the essential services that ecosystems provide are under threat. The decline of biodiversity undermines the resilience of natural systems and the well-being of human societies.

Addressing these challenges requires urgent action to reduce greenhouse gas emissions, protect and restore habitats, and manage land use sustainably. By safeguarding biodiversity, we also strengthen the planet's ability to cope with and mitigate the impacts of climate change, ensuring a more stable and resilient future for all life on Earth.

Summary

1. Climate change ecology examines how climate shifts affect organisms, ecosystems, and their interactions.
2. Human activities like burning fossil fuels, deforestation, and industrial agriculture are the main drivers of recent climate change.
3. The intensified greenhouse effect from gases like CO_2, CH_4, and N_2O leads to global warming and extreme weather events.
4. Climate influences species distribution, biological timing (like flowering and migration), and ecosystem productivity.
5. Many species are shifting to cooler regions, and mismatches in ecological timing are disrupting food chains.
6. Coral reefs are endangered by ocean warming and acidification, affecting marine biodiversity and fisheries.
7. Forests, wetlands, and oceans provide essential services like carbon storage and climate regulation but are being degraded.
8. Agriculture and coastal ecosystems are highly vulnerable to changing temperatures, rainfall, and rising sea levels.

9. Climate change intensifies other environmental stresses like habitat loss, pollution, and invasive species.
10. Mitigation includes reducing emissions and promoting renewable energy, while adaptation focuses on habitat restoration and resilience-building.
11. Nature-based solutions such as reforestation and wetland conservation benefit both the climate and biodiversity.
12. Natural climate factors (like solar cycles and volcanic eruptions) exist but recent warming is largely human-induced.
13. Feedback loops like melting ice, permafrost thaw, and ocean acidification worsen climate impacts.
14. Global temperatures have risen over 1°C since the 19th century, with more rapid warming in the Arctic and land areas.
15. Increased heatwaves, shifting rainfall, and seasonal changes affect both ecosystems and human life.
16. Biodiversity is under threat, with rising extinction risks and disruptions in species interactions and ecosystem stability.
17. Urgent global action is needed to reduce emissions, protect biodiversity, and support ecosystems for long-term sustainability.

Very Short Answer Type Questions

1. Define climate change ecology.
2. What is the greenhouse effect?
3. Name any two greenhouse gases.
4. What is ocean acidification?
5. Mention one impact of climate change on coral reefs.
6. Give an example of a nature-based solution.
7. What are Milankovitch cycles?
8. What is albedo?
9. Mention one human activity contributing to climate change.
10. What is black carbon?

Short Answer Type Questions

1. How does climate influence species distribution?
2. What are the effects of deforestation on climate?
3. Explain the role of volcanoes in climate change.
4. List two human-induced causes of climate change.

5. What is the difference between weather and climate?
6. State two observed ecological changes due to climate change.
7. What are feedback mechanisms? Give an example.
8. Why are Arctic regions more affected by global warming?
9. What is the role of coastal ecosystems in mitigating climate change?
10. How does climate change impact agricultural systems?

Long Answer Type Questions (5 marks each)

1. Describe the effects of climate change on biodiversity and ecosystem services.
2. Explain how climate change interacts with other environmental stressors.
3. Discuss the observed and projected changes in species distribution due to climate change.
4. Illustrate the human-induced factors contributing to climate change with examples.
5. How do ecosystems contribute to climate change mitigation?
6. Explain the changes in weather and temperature patterns caused by climate change.
7. Describe how extreme weather events affect ecosystems.
8. What are the impacts of climate change on marine ecosystems?
9. Elaborate on the disruption of biological cycles and species interactions due to climate change.
10. Suggest strategies for adapting to and mitigating the ecological impacts of climate change.

Multiple Choice Questions (MCQs)

1. What is the primary cause of recent global climate change?
   a. Volcanic eruptions
   b. Variations in solar radiation
   c. Human activities such as burning fossil fuels
   d. Milankovitch cycles
2. Which of the following gases is not a major greenhouse gas?
   a. Carbon dioxide
   b. Methane
   c. Oxygen
   d. Nitrous oxide
3. Ocean acidification is mainly caused by:
   a. Oil spills
   b. Plastic pollution
   c. Excess carbon dioxide absorption
   d. Industrial waste
4. Which of these ecosystems acts as a major carbon sink?
   a. Deserts
   b. Coral reefs
   c. Forests
   d. Urban areas

5. A key consequence of rising global temperatures is:
   a. Increased snowfall in all regions
   b. Uniform warming across the globe
   c. More frequent extreme weather events
   d. Permanent cooling of the oceans
6. The mismatch between plant flowering and pollinator activity is an example of:
   a. Ecosystem migration
   b. Phenological disruption
   c. Biodiversity enhancement
   d. Albedo effect
7. Which of the following is a nature-based solution to climate change?
   a. Use of fossil fuels
   b. Deforestation
   c. Reforestation
   d. Urban expansion
8. Which region is warming at more than twice the global average?
   a. Antarctic
   b. Equator
   c. Arctic
   d. Amazon Basin
9. The term "albedo" refers to:
   a. The acidity of ocean water
   b. The amount of heat trapped by greenhouse gases
   c. The reflectivity of Earth's surface
   d. The melting of polar ice
10. One major effect of climate change on agriculture is:
    a. Uniform increase in crop yields
    b. Enhanced soil fertility everywhere
    c. **Changes in rainfall patterns and reduced water availability**
    d. Elimination of pest attacks

11

Urban Ecology

Urban Expansion

The expansion of urban areas into surrounding regions, known as urban expansion, is driven by various factors including demographic shifts, economic growth, and societal changes. Comprehending current trends in urban expansion is crucial for tackling issues related to sustainability, infrastructure development, and environmental consequences. This phenomenon can result from rural-to-urban migration, leading to urban intensification. Disparities in birth and death rates, coupled with enhanced medical facilities in cities, may contribute to higher natural population growth in urban centers compared to rural areas. Urban expansion can occur through several mechanisms: the urbanization of previously rural land encircled by urban areas, the spatial growth of existing urban zones, and the emergence of new urban centers disconnected from established urban regions. As the global population has become predominantly urban, the scale and impact of urban expansion have intensified, with cities serving as primary hubs for residence and activity. The spatial manifestation of urban expansion is reflected in metropolitan growth, encompassing alterations in land use patterns, spatial enlargement, and changes in city dimensions within metropolitan regions.

Key Drivers of Urban Expansion

1. **Demographic Dynamics:** The influx of individuals to urban centers seeking improved employment prospects, educational opportunities, and amenities drives population growth. This increase necessitates urban expansion to accommodate rising numbers, significantly impacting economic systems, environmental conditions, and social structures worldwide. Population growth fuels demand for housing, services, and employment, leading to the physical spread of urban areas. Factors contributing to population growth include natality, mortality, and migration patterns. Larger urban populations can invigorate local economies, draw businesses, and generate employment opportunities, while also fostering diverse communities with a rich tapestry of cultures and ideas.

2. **Economic Progression:** The trajectory of economic development plays a pivotal role in urban expansion by generating employment opportunities, elevating income levels, and stimulating infrastructure investments. As cities evolve into economic nuclei, they attract migrants seeking job prospects, necessitating spatial growth. The increased demand for housing and services transforms agricultural or undeveloped land into urban settlements. The establishment of educational institutions, healthcare facilities, and recreational amenities supports elevated urban living standards and facilitates city expansion.
3. **Technological Innovation:** Advancements in transportation and communication technologies can enhance the accessibility and desirability of previously remote areas, promoting urban spread. Each phase of urban expansion is profoundly influenced by technological breakthroughs that reshape living, working, and mobility patterns. These advancements encompass developments in transportation systems, communication networks, construction methodologies, and energy resources.

Applications in Urban Expansion

1. The advent of the steam engine revolutionized travel and transportation modalities.
2. The widespread adoption of automobiles in the 20th century facilitated easier and more comfortable long-distance travel.
3. The transition from coal to oil and gas as primary energy sources supported industrial growth and urban development during and following the Industrial Revolution

Limitations of Urban Expansion

1. Urban expansion encompasses both beneficial and detrimental aspects. When implemented judiciously, it can minimize negative outcomes while maximizing advantages. The consequences of urban expansion include:
2. Inequitable resource distribution, housing scarcity, and increased urban poverty due to rapid growth.
3. Heightened strain on transportation infrastructure, water resources, energy systems, and waste management facilities.
4. Degradation of natural habitats, elevated pollution levels, and increased carbon emissions resulting from urban sprawl.

5. Urban development often overlooks equitable resource allocation, exacerbating wealth disparities.
6. Encroachment into natural areas results in habitat fragmentation and loss, impacting biodiversity and ecosystem function.
7. Expanding cities struggle to maintain efficient transportation networks, leading to traffic congestion and extended commute times.
8. The rising demand for land in growing urban areas inflates property values, rendering housing unaffordable for lower-income residents.

Urban Biodiversity

The global human population is experiencing exponential growth, with over 50% now residing in urban areas. This rapid expansion intensifies anthropogenic impacts, giving rise to numerous environmental challenges, including degraded air and water quality, soil erosion, diminished green spaces, and reduced natural habitats, including water sources. Paradoxically, these land-use alterations have also contributed to an increase in urban biodiversity, a phenomenon that has only recently gained attention. Historically, research efforts have primarily focused on conserving biodiversity in rural settings, operating under the assumption that urban habitats are of inferior environmental quality. Urban biodiversity encompasses the diverse array of living organisms within city environments, including flora, fauna, fungi, and microorganisms. It comprises the ecosystems and habitats found in urban settings, such as parks, gardens, waterways, and green roofs, and their interactions with the built environment and human activities.

Significance of Urban Biodiversity

Urban biodiversity holds considerable importance due to its provision of various ecosystem services, including enhanced air quality, climate regulation, and water management. Additionally, it offers health benefits, contributes aesthetic value, and fosters biodiversity conservation. These advantages can be classified into three primary categories:

1. Ecosystem Services
2. Human Well-being
3. Biodiversity Conservation

1. Ecosystem Services

1. Urban vegetation, particularly trees and plants, plays a vital role in maintaining air quality. They function as natural air purifiers, absorbing

pollutants such as nitrogen dioxide, sulfur dioxide, and particulate matter through their foliage. Furthermore, they generate oxygen and sequester carbon dioxide, creating a healthier environment for various organisms and substantially reducing pollution levels, resulting in cleaner air for urban inhabitants.

2. Natural urban landscapes, including parks, gardens, and wetlands, serve as effective rainwater absorption systems, mitigating surface runoff. This function helps prevent urban flooding and alleviates pressure on stormwater management infrastructure. Green roofs and permeable surfaces facilitate groundwater recharge by allowing water to infiltrate the soil, simultaneously filtering out contaminants and contributing to the overall health of urban waterways.
3. Urban vegetation plays a crucial role in mitigating the urban heat island phenomenon, wherein metropolitan areas experience substantially elevated temperatures compared to surrounding rural regions. Arboreal and herbaceous plants provide shade and engage in transpiration, releasing water vapor that cools the surrounding atmosphere. Additionally, urban green spaces function as carbon sinks, sequestering carbon dioxide and thus contributing to the mitigation of climate change. These areas also promote energy conservation by reducing ambient temperatures and subsequently decreasing the necessity for artificial cooling systems.

2. Human Well-being

1. Green space accessibility offers myriad health advantages. Engagement in physical activities such as ambulatory exercise, running, and participation in sports within park environments enhances cardiovascular function and mitigates obesity risk. Moreover, these verdant areas confer psychological benefits, alleviating stress, anxiety, and depressive symptoms. Exposure to natural environments has been empirically demonstrated to elevate mood, induce relaxation, and enhance overall well-being.
2. Urban parks and gardens serve as communal hubs, facilitating social interaction and recreational pursuits. These spaces foster community cohesion and social integration, providing venues for civic events, cultural activities, and interpersonal exchanges that reinforce neighborhood bonds and elevate residents' quality of life.
3. Biodiverse urban landscapes contribute significantly to the aesthetic appeal of cities. The presence of varied flora and fauna enhances visual interest and beauty, rendering urban environments more appealing and

livable. Well-maintained green spaces have been shown to augment property values, stimulate tourism, and engender a sense of civic pride and contentment among urban dwellers.

3. Biodiversity Conservation

1. Contrary to conventional wisdom, urban areas can function as vital wildlife refugia, particularly in the face of habitat fragmentation due to ongoing development. Urban green spaces can sustain a diverse array of species, encompassing avian, entomological, mammalian, and reptilian fauna. By offering sustenance, shelter, and breeding grounds, cities can contribute to the maintenance of biodiversity and the creation of vibrant urban ecosystems.
2. Urban biodiversity initiatives play a pivotal role in the preservation of endangered species and genetic diversity. Botanical gardens, urban wildlife sanctuaries, and conservation programs serve as safeguards for rare and threatened species, ensuring their continued existence. The integration of biodiversity considerations into urban planning and development enables cities to contribute to global conservation efforts and establish sustainable habitats for future generations.

Challenges to Urban Biodiversity

1. Urban expansion frequently results in the fragmentation of natural habitats, impeding species mobility and resource acquisition.
2. Pollution of air, water, and soil, stemming from industrial activities, transportation, and waste management, can adversely affect urban biodiversity.
3. The introduction of non-native species to urban environments can lead to competition with indigenous species and disruption of local ecosystems.
4. Urban areas exhibit heightened vulnerability to climate change impacts, including temperature increases and extreme weather events, which can place significant stress on urban ecosystems.
5. The intense demand for land in urban centers often constrains the availability of green spaces and habitats essential for biodiversity.

Outcome of Urban Expansion and Biodiversity on Natural Habitats

The process of urbanization and its associated land cover transformation engenders substantial environmental modifications, driven by population growth, economic dynamics, lifestyle preferences, and the demand for built-up areas with infrastructure. This results in high-density land utilization and the alteration of natural ecosystems. Land cover change, a fundamental aspect of urbanization, overlaps with the intensification of infilling. The expansion of urban areas occurs at the cost of valuable natural habitats, presenting unique management challenges. Globally, natural ecosystems are vanishing at an alarming pace, with remaining ecosystems facing significant anthropogenic pressures. The patterns of land cover conversion significantly influence growth trajectories, socio-economic conditions, and societal prosperity, particularly in emerging economies and developing nations. These alterations in land cover have resulted in considerable losses in upstream and downstream hydrological flow potential, affecting discharge, modifying storm hydrographs, increasing sediment transport, and transforming typical ecosystem biogeochemical and hydrologic functions.

Consequences of Urban Expansion on Biodiversity

1. The direct loss of natural habitats due to urban expansion, as land is repurposed for residential, infrastructural, and industrial development, leads to the destruction of ecosystems crucial for various species.
2. Urban environments tend to favor generalist species capable of thriving in diverse conditions, while specialist species requiring specific habitats and resources often experience decline.
3. Habitat fragmentation occurs as continuous natural areas are subdivided into smaller, isolated patches, impeding wildlife movement, foraging, and reproduction.
4. The expansion of urban areas can disrupt essential ecosystem processes, including pollination, seed dispersal, and nutrient cycling.
5. Habitat loss and fragmentation can result in biodiversity decline, as numerous species struggle to adapt to environmental changes or locate suitable alternative habitats.
6. Urban-generated pollution, encompassing air, water, and soil contamination, further degrades natural habitats and adversely affects wildlife populations.
7. Urban centers can serve as introduction points for invasive species, which may outcompete native species and alter ecosystem dynamics. These non-native species can be introduced through various human activities, including landscaping and transportation.
8. Urban areas contribute to climate change through greenhouse gas emissions and generate urban heat island effects, where cities experience significantly higher temperatures than surrounding rural areas, further stressing natural habitats and species.

Implement Strategies to Enhance Urban Biodiversity

1. Integrating green roofs, living walls, and porous surfaces into urban design to enhance biodiversity and deliver ecosystem services.
2. Employing indigenous flora in urban landscaping to sustain local fauna.
3. Facilitating resident participation in biodiversity preservation efforts, including community horticulture, citizen-led scientific initiatives, and ecosystem restoration projects.
4. Enacting policies and regulations to safeguard and augment urban biodiversity, encompassing green space conservation, pollution mitigation, and invasive species control.
5. Installing artificial habitats such as avian shelters, chiropteran roosts, and entomological refuges to support diverse species.
6. Integrating elements like rain gardens, bioswales, and engineered wetlands to manage stormwater runoff and foster biodiversity.
7. Establishing green corridors to link isolated natural areas, facilitating safe wildlife movement across urban landscapes.
8. Utilizing railway corridors and roadside verges for indigenous vegetation planting to create linear ecological habitats.
9. Launching educational programs in academic institutions and communities to heighten awareness regarding biodiversity significance and conservation strategies.

Case Study: Urban Expansion's Impact on Biodiversity in Chandaka Wildlife Sanctuary, Odisha

The expansion of urban areas adjacent to Chandaka Wildlife Sanctuary has resulted in significant habitat fragmentation and loss, profoundly affecting its biodiversity. The sanctuary's previously uninterrupted habitats have been segmented due to urban land transformation. The incorporation of the Bharatpur Range Forest, a component of the sanctuary, into Bhubaneswar's municipal boundaries exemplifies habitat encroachment. This fragmentation disrupts ecological equilibrium, hindering species' capacity to locate sustenance, shelter, and mates, consequently leading to population declines. Elephants in Chandaka have been notably impacted, given their requirement for extensive, undisturbed territories. Encroachment has precipitated human-elephant conflicts, resulting in damage to property and crops, jeopardizing both livelihoods and conservation endeavors. The sanctuary's biodiversity is further imperiled by elevated pollution levels and altered hydrological regimes stemming from urban runoff and waste disposal practices. Investigators have utilized satellite imagery and indices such as the Normalized Difference Vegetation Index (NDVI) and Normalized Difference Built-up Index (NDBI) to monitor land use alterations and their impact on biodiversity, revealing rapid spatial transformations over the past two decades attributable to urban development. To ameliorate these impacts, recommendations include implementing zoning regulations to manage urban growth while preserving habitats, enhancing the Chandaka Wildlife Management Action Plan, and fostering community engagement and awareness for sustainable development. Involving local communities can cultivate environmental stewardship and balance development with ecological conservation. Despite the challenges, Chandaka Wildlife Sanctuary demonstrates the potential for integrating conservation principles into urban planning frameworks. A comprehensive approach incorporating stringent zoning laws, refined management strategies, and active community participation can mitigate the adverse effects of urbanization and promote biodiversity conservation.

Very Short Answer Types Questions

1. What are the main drivers of urban expansion?
2. What are the primary factors contributing to population growth?
3. Describe urban expansion?
4. What are ecosystem services?
5. List out any two challenges faced due to urban biodiversity.
6. Explain how population growth can lead to challenges in urban infrastructure.
7. Discuss any 2 applications of urban expansion.
8. What strategies can implement to enhance urban biodiversity?

Short Answer Type Questions

1. Describe how economic development can lead to population growth?
2. What is urban biodiversity? Explain its significance.
3. How technological innovation promotes urban expansion? What are the pros and cons of it?
4. What are the major challenges faced in urban biodiversity?
5. Why does increased income tends to accelerate urban growth?
6. Explain how industrialization acts as a catalyst for urban expansion with an example.
7. Name two communication technologies that have impacted urban connectivity.

Long Answer Type Questions

1. Discuss the limitations of urban expansion.
2. Discuss the significance of urban biodiversity in detail.
3. What are the negatives impact of urban expansion and biodiversity on natural habitats.
4. What key steps can be taken to enhance urban biodiversity?
5. Discuss the outcomes of urban expansion and biodiversity on natural habitats.
6. Describe a case study that discuss the outcomes of urban expansion on biodiversity.
7. Describe the challenges of urban biodiversity.
8. Discuss the consequences of urban expansion?

Multiple Choice Questions

1. The factors affecting population growth are-
   a. Birth rate
   b. Death rate
   c. Migration
   d. All of them.
2. The drivers of urban expansion are-
   a. Economic development
   b. Population growth
   c. Technological innovation
   d. Only i and iii
   e. All of them
3. Major reasons of urban expansion
   a. Increased population growth
   b. Advanced technology demands
   d. Constructing parks, forest.
   c. Both i and ii
4. The significance of urban biodiversity is
   a. It prevents flooding and absorb harmful pollutants from the atmosphere.
   b. Promotes health advantages
   c. Increase nitrogen dioxide and other particulate in atmosphere
   d. None of them
5. Strategies to enhance urban biodiversity are
   a. Habitat destruction
   b. Deforestation
   c. Habitat fragmentation
   d. Integrate green roofs, live walls to enhance biodiversity

6. NDBI stands for
   a. Normalized Difference building index
   b. Normalized Difference Built-up Index
   c. Normalized Difference Built infrastructure
   d. None of the above
7. NDVI stands for

a. Normalized Difference Vegetation Index

b. Normal Difference Variation index

c. Normalized Different Variation Index

d. None of the above

Section V
Landscape, Conservation and Sustainable Practices

12

Landscape Ecology

Introduction

Landscape ecology is a branch of ecological science that focuses on the patterns and interactions among ecosystems within a given region, emphasizing how spatial arrangement and heterogeneity influence ecological processes. Unlike traditional ecology, which often examines isolated ecosystems or populations, landscape ecology explicitly considers the mosaic of different habitat types, land uses, and physical features that make up real-world environments. This approach is essential for understanding the complexity of nature and the effects of human activities on the environment.

The Scope and Definition of Landscape Ecology

At its core, landscape ecology examines the spatial variation within landscapes—how different ecosystems, such as forests, wetlands, grasslands, agricultural fields, and urban areas, are distributed and how their arrangement affects the flow of energy, materials, and organisms. A landscape is not defined solely by its size but by its composition (the types and abundance of ecosystems present), configuration (the spatial arrangement and connectivity of these elements), and the processes occurring within and between them.

Landscape ecology is inherently interdisciplinary, integrating concepts from ecology, geography, planning, and the social sciences. This integration allows for the study of both natural and human-modified landscapes, recognizing that human activities are a major force shaping the structure and function of modern landscapes.

Historical Background

The origins of landscape ecology can be traced back to the early 20th century, with the term first introduced by German geographer Carl Troll in 1939. Troll emphasized the use of aerial photography to study the relationships between environment and vegetation, laying the groundwork for the spatially explicit perspective that defines the field today. However, landscape ecology truly gained prominence in the 1980s, as ecologists and geographers sought to address ecological questions at broader spatial scales and to understand the impacts of rapid environmental change, habitat fragmentation, and land-use transformation.

Key Concepts in Landscape Ecology

Spatial Heterogeneity: A fundamental concept in landscape ecology is heterogeneity—the idea that landscapes are composed of a diversity of elements that vary in type, size, shape, and distribution. This spatial diversity influences ecological processes such as species movement, nutrient cycling, and energy flow.

Composition, Configuration, and Connectivity

1. *Composition* refers to the variety and proportion of different habitat types or land covers within a landscape.
2. *Configuration* is the spatial arrangement and organization of these elements, including the size and shape of patches, the presence of corridors, and the degree of fragmentation.
3. *Connectivity* describes how easily organisms, energy, and materials can move between different parts of the landscape. High connectivity facilitates movement and ecological resilience, while low connectivity can lead to isolation and reduced biodiversity.

Fragmentation and Matrix

Fragmentation occurs when continuous habitats are broken into smaller, isolated patches, often due to human activities such as agriculture, urbanization, or infrastructure development. The *matrix* is the dominant background land cover in which these patches are embedded, influencing the movement and survival of species.

Scale and Pattern

Landscape ecology places a strong emphasis on scale—both spatial (from local to regional to global) and temporal (from days to decades or centuries). Patterns observed at one scale may not be apparent at another, and ecological processes can operate differently depending on the scale of observation.

The Interdisciplinary Nature of Landscape Ecology

Landscape ecology bridges the gap between natural and social sciences, incorporating biophysical, analytical, and humanistic perspectives. It recognizes that landscapes are shaped by both natural forces (such as climate, geology, and hydrology) and human activities (such as land use, resource management, and urban development). This holistic view is essential for understanding how landscapes function and for developing strategies to manage and conserve them effectively.

Research Themes and Applications

Landscape ecology addresses a wide range of research topics, including:

1. The movement of species, water, and nutrients across landscape mosaics
2. The effects of land use and land cover change on ecosystem health
3. The relationship between spatial pattern and ecological process
4. The conservation and restoration of habitats and biodiversity
5. The management of ecosystem services that support human well-being

The practical applications of landscape ecology are vast. It informs land-use planning, conservation strategies, and environmental management by providing insights into how spatial patterns affect ecological processes and how human activities can be integrated sustainably within the natural world.

Human Influence and Contemporary Relevance

In the modern era, landscapes are increasingly shaped by human activities, leading to challenges such as habitat loss, fragmentation, pollution, climate change, and the spread of invasive species. Landscape ecologists study how these changes alter the structure and function of landscapes and develop strategies to mitigate negative impacts, restore degraded areas, and promote sustainability.

The field is particularly relevant for managing the interface between natural and human-dominated systems, recognizing that ecological processes do not stop at the boundaries of protected areas. Instead, landscape ecology encourages a comprehensive view that considers the entire mosaic of land uses and the flows of organisms, energy, and materials across different landscape elements.

Summary

1. Landscape ecology studies spatial patterns and interactions among ecosystems across a region.
2. It emphasizes heterogeneity and the influence of spatial arrangement on ecological processes.
3. Key components include **composition**, **configuration**, and **connectivity** of ecosystems.

4. A **landscape** includes various habitats (e.g., forests, wetlands, urban areas) and is defined by structure, not size.
5. The field is **interdisciplinary**, involving ecology, geography, planning, and social sciences.
6. **Fragmentation** breaks continuous habitats into patches, often due to human activity; the **matrix** is the surrounding land cover.
7. Concepts of **scale** (spatial and temporal) are vital; patterns/processes vary across different scales.
8. The discipline addresses **species movement, land-use changes, ecosystem services, and conservation**.
9. Human influences—urbanization, agriculture, pollution—are key drivers of landscape transformation.
10. Landscape ecology aids in **planning, habitat restoration, biodiversity conservation, and sustainability**.

Very Short Answer Type Questions

1. Who coined the term “landscape ecology”?
2. Define landscape ecology.
3. What is meant by “composition” in landscape ecology?
4. Name one human activity that causes fragmentation.
5. What is the “matrix” in a landscape?
6. Mention one key goal of landscape ecology.
7. What tool did Carl Troll use in early landscape studies?
8. What does “connectivity” refer to in landscapes?
9. Name one element included in the configuration of a landscape.
10. What is spatial heterogeneity?

Short Answer Type Questions

1. Explain the importance of spatial arrangement in landscape ecology.
2. How does habitat fragmentation affect biodiversity?
3. Briefly describe the role of the matrix in landscape ecology.
4. Distinguish between composition and configuration.
5. What are the main drivers of landscape change in modern times?
6. Explain the interdisciplinary nature of landscape ecology.
7. Why is the concept of scale important in landscape studies?
8. List two applications of landscape ecology in environmental planning.

9. How can landscape ecology help in conserving ecosystem services?
10. Describe the relationship between spatial patterns and ecological processes.

Long Answer Type Questions

1. Describe the key concepts of landscape ecology with examples.
2. Discuss the historical development of landscape ecology.
3. Explain the effects of fragmentation and the role of the matrix.
4. Elaborate on how landscape ecology integrates natural and social sciences.
5. Analyze the importance of composition, configuration, and connectivity in landscape structure.
6. How does landscape ecology contribute to biodiversity conservation and sustainable land use?
7. Discuss the relevance of scale (spatial and temporal) in understanding landscape patterns.
8. Examine the role of landscape ecology in addressing human-induced environmental challenges.
9. How can landscape ecology support habitat restoration and planning efforts?
10. Describe major research themes in landscape ecology and their real-world applications.

13

Conservation Genomics

Conservation genomics, a burgeoning discipline, employs genomic techniques to support the conservation and viability of diverse species and populations. This field examines an organism's complete genome to comprehend genetic diversity, population dynamics, and evolutionary mechanisms. The main objectives of conservation genomics include evaluating genetic variation within and among populations, which is essential for maintaining genetic diversity and ensuring species' long-term survival. It offers valuable insights into population size, gene flow, and genetic drift, all of which are critical for developing effective conservation strategies. Genomic information can assist in identifying species at risk of extinction and prioritizing conservation efforts. By elucidating the genetic foundations of traits that contribute to survival and reproduction, conservation genomics can inform the development of targeted conservation approaches. It also aids in investigating hybridization events and species' adaptations to changing environments, whichiscrucialformanagingand preserving biodiversity. As a powerful tool, conservation genomics enhances traditional conservation genetics by providing a more comprehensive understanding of the genetic factors influencing species survival and ecosystem health.

What is Genetic Diversity?

Genetic diversity refers to the diversity of alleles and genotypes present in apopulation, species, or species group. It is manifested through numerous traits such as pigmentation, enzyme composition, and DNA sequences in various organisms. Such diversitymayexpress genomic variations subtly and manifests itself in the form of protein sequences of aminoacids from particular loci. Protein variation may lead to functional differences impacting individual reproduction, survival, or behavior. The capacity for populations to adapt to environmental shifts hinges on genetic diversity, which is typically abundant in large, naturally out breeding populations but often diminished in endangered species. Notably, genetic diversity among developmental loci can influence phenotypic outcomes, as exemplified by the wide range of dog breeds.

The Significance of Genetic Diversity

Genetic diversity is central to ensuring the continued health and survival of species. The greater adaptive responses it supports provide robustness against a variety of environmental pressures - from changing climates to earthquakes and epizootics. Ultimately, genetic diversity provides a raw material on which to base generational changes to keep species alive indefinitely. The more diverse a population's genetic makeup, the higher the chances of possessing individuals that are resistant to diseases, thereby reducing the chances of the entire population being destroyed by a pathogen. Genetic diversity among species contributes to the strength of an ecosystem, with each species playing its functional role and maintaining stability in an ecosystem.

Genetic diversity among crops and livestock ensures food security through the development of crop varieties that are resistant to pests, diseases, and climatic changes. Conservation efforts therefore are targeted on genetic diversity for maintaining the adaptive potential of species and biodiversity for the generations yet to come. Three reasons for conserving genetic diversity have been underlined by the International Union for Conservation of Nature, according to McNeely et al. (1990), including its necessity for evolutionary adaptation, its role in avoiding inbreeding depression and reduced fitness in outbreeding species, and its contribution to ecosystem diversity. Studies have proven that genetic diversity in plants is linked to the species diversity of phytophagous and enhances resilience in the ecosystem against perturbations (Reusch et al. 2005; Crutsinger et al. 2006).

Extent of Genetic Diversity

Outbreeding species with large populations generally possess extensive genetic diversity. This diversity is evident in the wide range of morphological, behavioral, and physiological variations observed within most populations. Such variation stems from both environmental factors influencing individuals (non-genetic-based variation) and differences in alleles and heterozygosity across numerous loci (genetic-based variation). The diverse array of dog breeds derived from the wolf genome exemplifies the substantial genetic diversity inherentin a species. Barring a few mutations, the variety of dog breeds reflects the genetic diversity present in their wolf ancestors. Genetic diversity can be assessed at variouslevels, including quantitative variation in measurable traits, visible effects of harmful alleles, protein variations, and direct analysis of DNA sequence differences.

Factors Affecting Genetic Diversity

1. **Mutation:** Alterations in DNA sequences, known as mutations, generate new genetic variants. These can arise from errors in DNA replication or exposure to environmental agents like radiation or chemicals. As the primary source of all genetic variability, mutations can introduce advantageous, neutral, or detrimental changes to an organism's genetic makeup.
2. **Gene Flow:** Gene flow, also termed gene migration, involves the exchange of genetic material between different populations through interbreeding. This transfer of genes enhances genetic variability by introducing novel alleles (gene variants). An example of gene flow is the fertilization of one plant population by pollen from another.
3. **Genetic Drift:** Genetic drift refers to random fluctuations in allele frequencies within a population.This phenomenon is particularly significant in small populations, where chance events can substantially impact genetic variability. For example, a natural disaster that drastically reduces population size may result in surviving individuals that do not fully represent the genetic diversity of the original group.
4. **Natural Selection:** Natural selection is the process whereby organisms better suited to their environment have a higher likelihood of survival and reproduction. This selective pressure can increase the prevalence of beneficial traits while decreasing the frequency of harmful ones. Over time, natural selection molds the genetic variability of a population by favoring advantageous genetic variants.
5. **Population Size:** The number of individuals in a population influences its genetic variability. Larger populations tend to maintain more genetic variation due to their reduced susceptibility to genetic drift and in breeding. Conversely, smaller populations are more prone to losing genetic variability through random events and in breeding, potentially leading to a higher incidence of genetic disorders.
6. **Reproductive Strategies:** The method of reproduction affects genetic variability. Sexual reproduction mixes genes through meiosis and fertilization, creating new allele combinations and increasing genetic variability. In contrast, asexual reproduction produces genetically identical offspring, resulting in lower genetic variability.
7. **Environmental Factors:** Environmental conditions such as climate, food availability, and habitat characteristics can exert selective pressures on a population. These pressures influence which genetic traits are

advantageous, leading to shifts in allele frequencies over time. For instance, in a colder climate, individuals with traits that enhance heat retention may have a higher likelihood of survival and reproduction.

8. **Human Activities:** Human actions, including habitat destruction, pollution, selective breeding, and the introduction of non-native species, can significantly affect genetic variability. For example, habitat fragmentation can isolate populations, reducing gene flow and increasing in breeding risk. Conversely, conservation efforts and breeding programs can help maintain or enhance genetic variability.

These factors interact in intricate ways to shape the genetic variability of populations and species. Genetic variability is crucial for the health, adaptability, and survival of species, making its study and preservation a key focus in biology.

Conservation Strategies for Preserving Genetic Diversity

Maintaining genetic diversity is essential for species' health, adaptability, and longevity.The following are key approaches used in conservation efforts:

1. Wildlife Reserves and Protected Areas

1. Creating protected zones and wildlife sanctuaries aids in preserving habitats and sustaining natural populations, thus safeguarding diverse genetic traits.

2. Habitat Restoration

1. Rehabilitating degraded ecosystems can enhance conditions for species toflourish,thereby supporting genetic diversity. This includes activities such as reforestation, wetland rehabilitation, and management of invasive species.

3. Ex Situ Conservation

1. This method involves preserving genetic diversity outside natural habitats. Techniques encompass seed banks, botanical gardens, zoos, and aquariums where species are cultivated and researched for potential reintroduction into their natural environments.

4. Genetic Rescue

1. Integrating individuals from diverse populations into small, inbred groups can enhance genetic diversity and mitigate the risk of genetic disorders. This approach is employed to improve genetic health and resilience.

5. Legal and Policy Frameworks

Implementing and enforcing legislation and policies that safeguard endangered species their habitats is vital. International agreements, such as the Convention on Biological Diversity, bolster these efforts.

6. Sustainable Land use Practices

Encouraging sustainable methods inagriculture, forestry, and fishing helps conserve natural habitats and reduce impacts on biodiversity. These practices include crop rotation, agroforestry, and sustainable harvesting techniques.

7. Public Education and Awareness

1. Increasing public understanding of the significance of genetic diversity and biodiversity conservation fosters community support and involvement in conservation initiatives.

8. Conservation Breeding Programs

Captive breeding programs can help maintain genetic diversity and provide a safeguard against extinction. These programs often involve meticulous genetic management to ensure diverse and healthy offspring.

9. Monitoring and Research

Ongoing monitoring and research help track genetic diversity levels, evaluate the efficacy of conservation strategies, and modify approaches based on new scientific insights.

10. Climate Change Mitigation

Addressing climate change through mitigation efforts reduces its impact on species and their habitats, helping to preserve genetic diversity. Strategies include reducing greenhouse gas emissions and protecting carbon sinks.

Genomic Technologies to conserve Biodiversity

Biodiversity conservation is being transformed by genomic technologies, which offer potent instruments for examining and safe guarding genetic diversity The following are some crucial genomic technologies and their conservation applications:

1. Whole Genome Sequencing(WGS)

1. WG Sentails decoding an organism's entire DNA sequence.
2. It aids in recognizing genetic variations within and among species, evaluating genetic diversity, and uncovering adaptive traits that can support conservation endeavors.

2. GenomicSelection

1. This method utilizes genome-wide information to forecast the genetic worth ofindividuals for particular traits.
2. It is employed in breeding programs to choose individuals with sought after characteristics, such as disease resistance or drought tolerance, improving the genetic well-being of populations.

3. CRISPR-Cas9 GeneEditing

1. CRISPR-Cas9 is a precise genetic modification tool capable of making targetedalterations to living organisms' DNA.
2. It can be utilized to introduce beneficial genetic traits into endangered species or eliminate harmful mutations, potentially aiding in the recovery of threatened populations.

4. Environmental DNA (eDNA)

1. eDNA involves gathering and examining DNA from environmentals amples (e.g., soil, water) to detect species presence.
2. It isused to monitor biodiversity, identify rare or elusive species, and assess ecosystem health without direct observation.

5. Population Genomics

1. Population genomics investigates genetic variation within and between populations to comprehend evolutionary processes.
2. It assists in identifying population structure, gene flow, and adaptive genetic variation, informing conservation strategies like habitat connectivity and management of small populations.

6. Gene Drives

1. Gene drives are genetic systems that enhance the likelihood of a specific gene being transmitted to offspring.
2. They can be employed to control invasive species or vector-borne diseases by propagating genes that diminish the reproductive success of target species.

7. Metagenomics

1. Metagenomics involves sequencing DNA from environmental samples to study the collective genomes of microbial communities.
2. It helps understand microorganisms' roles in ecosystems, their interactions with host species, and their impact on biodiversity.

8. Conservation Genomics

1. Conservation genomics combines genomic data with conservation biology to address biodiversity loss.
2. It aids inidentifying genetic hotspots for protection, developing predictive models for species resilience, and informing management actions to mitigate threats.

These genomic technologies provide innovative approaches to conserving biodiversityby enhancing our understanding of genetic diversity, pinpointing conservation priorities, and developing targeted interventions. They represent a promising frontier in the effort topreserve our planet's rich biological heritage.

Case Studies to Understand Poaching

1. Pangolin Poaching in India: A Critical Conservation Challenge

The illegal hunting of pangolins in India represents a significant conservation concern, as these mammals are the most heavily trafficked worldwide. Two pangolin species inhabit India: the Indian pangolin (*Manis crassicaudata*) and the Chinese pangolin (*Manis pentadactyla*). These animals are extensively hunted for their scales, which are utilized in traditional medicine, and their meat, considered a delicacy in certain cultures. A study conducted by the Wildlife Conservation Society-India recorded 522 distinct wild life poaching and trafficking events in 2020, with pangolins among the most frequently targeted species. The illicit trade of pangolins is fueled by strong demand, both within India and internationally. Despite strict legislation and law enforcement efforts, the illegal pangolin trade continues to flourish. The study emphasized the necessity for improved enforcementand increased public awareness to combat this illicit trade. Obstacles include the extensive areas requiring monitoring, sophisticated poaching techniques, and low conviction rates for poaching-related offenses. The research identified primary trade routes and stressed the importance of accurate forensic procedures to trace trade networks. From 2009 to 2018, an estimated 7,500 pangolins were confiscated in 119 seizures. The increasing frequency of pangolin seizures over time suggests rising levels of poaching and trade.

Efforts to counter pangolin poaching include enhanced surveillance methods, such asdrone and camera trap deployment, and increased funding for habitat protection. Conservation initiatives also focus on community engagement and eco-tourism projects to mitigate human-wildlife conflict and promote sustainable livelihoods. Raising public awareness about pangolin importance and the threats they face is crucial for reducing demand for pangolin products. Educational campaigns and community involvement can help shift public attitudes and decrease the demand for illegal wildlife products. Pangolin poaching is a transnational issue, necessitating international cooperation for effective conservation. Collaborative efforts between countries can aid in tracking and dismantling illegal trade networks and strengthening enforcement measures.

In conclusion, pangolin poaching in India presents a significant challenge that requiresa multifaceted approach, including enhanced enforcement, public awareness campaigns, and international collaboration. By addressing these issues, we can work towards conserving pangolin populations and ensuring their long-term survival.

2. Tiger Poaching in India

Historically, India has been a habitat for a substantial tiger population; however, poaching has dramatically diminished their numbers. Throughout the British Rajeraand post-independence period, tigers were hunted for their pelts and various body parts. The population plummeted from an estimated 100,000 at the onset of the 20th century to approximately 3,000 in contemporary times. Despite conservation initiatives, poaching continues to pose a significant threat. Tiger components, including pelts, claws, osseous tissue, and dentition, are highly prized in illicit markets, particularly in China, where the yare employed in traditional medicine and as status symbols. Regions such as Maharashtra, Karnataka, Tamil Nadu, and Assam serve as epicenters for poaching activities. The Indian government has implemented various measures to combat this issue, including the Wildlife Protection Act of 1972 and Project Tiger. Nevertheless, enforcement remains challenging due to the vast territories requiring surveillance and the sophisticated techniques employed by poachers. The conviction rate for poaching-related offenses remains low, with numerous cases going unreported or unresolved.

In December 2024, authorities in Madhya Pradesh apprehended three individualsforpoaching a tiger in the Banspani forest area within the Narmadapuram forest division. The perpetrators had deployed live electrical wiring in the forest, resulting in the tiger's demisevia electrocution. The Sariska Tiger Reserve incident of 2005 revealed that the entire tiger population had been eradicated due to poaching activities. This occurrence underscored the imperative for enhanced monitoring and enforcement measures.

3. Bear Bile Trade in China

Bear bile has been utilized in Traditional Chinese Medicine (TCM) formillennia, purportedly to address various ailments, including hepatic disorders and pain management. The demand for bear bile has led to the establishment of bear farms, where the substanceis extracted from live specimens, primarily Asiatic black bears. While bear farming is legally sanctioned in China, the illicit trade in wild bear bile persists. Although legal farms are subject to regulation, concerns persist regarding animal welfare and the impact on wild populations. Illegal trade often involves the poaching of wild bears, posing a significant threat to their survival. Research investigating the bear bile market has revealed that a substantialproportion of consumers are unaware of the ethical implications of bile extraction. Many consumers perceive farmed bile as a more ethical alternative; however, both farmed and wild bile contribute to the overall demand for bear products. Reducing the demand for bear bile presents challenges due to its deep-rooted cultural significance and perceived medicinal properties. Efforts to mitigate demand include public awareness campaigns, promotion of alternative remedies, and more stringent enforcement of wildlife protection legislation. Conservation organizations are workingto safeguard wild bear populations by advocating for the closure of bear farms and promoting the use of synthetic alternatives to bear bile. International cooperation is crucial in combating illegal trade and protecting bear habitats.

Very Short Answer type Questions

1. Describe the meaning of the word poaching?
2. Discuss the purpose of conservation genomics.
3. Define genetic diversity
4. Explain the meaning of the term gene flow.
5. What is geneticdrift?
6. Tell the significance of Genetic Diversity.
7. Name the factors influencing genetic diversity.
8. What is Ex-Situ conservation?

Short Question Answer Type Questions

1. Describe genetic diversity and its importance with example.
2. Discuss Environmental factors that influence genetic diversity.
3. What are the key human activities responsible for genetic diversity?
4. Discussany two conservation strategies to preserve genetic diversity.
5. Discussany two genomic technologies that can help inconserving genetic biodiversity.
6. Discuss CRISPR-Cas9 Gene Editing.
7. What are factors influencing genetic variability?
8. Discuss the case studies of poaching.

Long Answer Type Questions

1. How does conservation of genomic data of widespread animal and plant species help?
2. Describe genetic diversity in detail along with examples and its significance.
3. Elaborate all the factors influencing genetic diversity.
4. What are the conservation strategies that can be employed top reserve genetic diversity.
5. Differentiate between conservation genomics and metagenomics.
6. Discuss some case studies of poaching inIndia.
7. Describe what happened during the Bear Bile Trade in China.

Multiple Choice Question

1. The method of preserving genetic diversity outside natural habitats is knownas
   - a. In situconservation
   - **b. Ex-situ conservation**
   - c. None ofthem
   - d. Both ofthem
2. Exchange of genetic material between different populations through interbreedingis known as
   - a. Gene migration
   - b. Geneflow
   - **c. Both**
   - d. None of the above
3. IUCN standsfor
   - **a. International Union for Conservation of Nature**
   - b. Indian Union forConservation
   - c. Integrated Universe for Methane
   - d. None of theabove
4. WGS standsfor
   - **a. Whole Genome Sequencing**
   - b. Whole Gene Sequence
   - c. Whole Genome Signing
   - d. None of the above

5. What is the full form of CRISPR-Cas9...............
6. Name the technique used for analysis of environmental samples to study the collective genome ....
7. What is the scientific name of a Chinese pangolin.............
8. What is *Manis crassicaudata*................
9. What are the factors of genetic diversity

   a. Mutation
   b. Genemigration
   c. **Both a and b**
   d. Geneticrescue

10. Which conservation strategies can be employed to control disease caused by invasive species

    a. Population genomics
    b. Gene Drives
    c. Metagenomics
    d. Gene mutations

14

Community-based Conservation

Local Communities and Biodiversity Conservation

Local communities have emerged as crucial stakeholders in biodiversity conservation efforts across the world. As the custodians of natural resources and traditional knowledge, they play a vital role in protecting ecosystems, species, and habitats that are often located in or near their settlements. In contrast to top-down approaches led solely by governments or external agencies, community-based conservation emphasizes the participation, ownership, and empowerment of local people in managing natural resources.

Many biodiversity-rich regions overlap with areas inhabited by indigenous populations and rural communities. These communities depend directly on forests, grasslands, wetlands, and marine environments for their livelihood, cultural practices, and spiritual beliefs. This close relationship with nature makes them both vulnerable to biodiversity loss and uniquely positioned to conserve it. Community-based conservation seeks to align conservation goals with the social, economic, and cultural interests of local people, thereby creating mutual benefits.

Empowering communities to manage their own resources often leads to better conservation outcomes. Traditional knowledge systems—such as indigenous farming methods, water harvesting practices, and sacred groves—have historically contributed to biodiversity maintenance. Recognizing, documenting, and integrating this knowledge into formal conservation strategies not only validates cultural heritage but also enhances ecological effectiveness.

Participation in conservation initiatives gives local people a sense of ownership and responsibility, which often results in more sustainable practices. Instead of being seen as threats or passive recipients of conservation policies, communities become active partners. For example, when local people are involved in patrolling forests or managing protected areas, they are more likely to deter illegal activities such as poaching, logging, or mining.

Community-led conservation is also more adaptive and context-specific. Local residents understand the ecological patterns, seasonal rhythms, and environmental changes better than outsiders. Their insights can guide decisions on land use, grazing, harvesting, and habitat restoration in a way that balances ecological health with human needs.

In some cases, communities have established their own conservation areas, such as community forests, fish sanctuaries, or heritage zones, where they enforce regulations and monitor biodiversity. These initiatives often outperform externally managed conservation projects because they are rooted in local governance and social norms. Incentive mechanisms such as eco-tourism, sustainable harvesting, and payment for ecosystem services can further strengthen community commitment to conservation.

However, successful community-based conservation requires supportive legal frameworks, transparent benefit-sharing, capacity-building, and genuine partnerships between communities and external stakeholders. Without these, projects risk becoming symbolic or exploitative, leading to resistance or failure. Land tenure security, access to decision-making platforms, and recognition of community rights are essential for long-term success.

In conclusion, local communities are not just passive inhabitants of biodiversity hotspots but active stewards whose involvement can transform conservation efforts. By building trust, respecting traditional knowledge, and ensuring equitable benefits, community-based conservation creates a model that is both ecologically sound and socially just.

Ecological and Social Sustainability

Ecological and social sustainability are foundational pillars of successful community-based conservation. Both are deeply interconnected, as long-term conservation outcomes depend not only on maintaining biodiversity and ecosystem integrity but also on ensuring that human communities thrive alongside nature. Without social sustainability, conservation efforts may face opposition, and without ecological sustainability, the natural resource base on which communities depend will degrade.

Ecological sustainability refers to the capacity of ecosystems to maintain their structure, function, and biodiversity over time, despite external pressures such as exploitation, pollution, or climate change. It involves protecting key species, habitats, and ecological processes—such as pollination, nutrient cycling, and hydrological flows—that support life. In the context of community-based conservation, ecological sustainability means managing resources in ways that do not compromise the ability of future generations to meet their needs.

Community-led monitoring, habitat restoration, rotational harvesting, and seasonal bans on resource use are practices that can promote ecological sustainability. For example, limiting fishing during breeding seasons or controlling grazing intensity helps ecosystems recover and maintain productivity. Communities that observe traditional ecological norms often intuitively promote such sustainability through cultural taboos and customary laws.

Social sustainability, on the other hand, focuses on the well-being, equity, and resilience of human communities. It ensures that conservation does not come at the expense of people's rights, livelihoods, or cultural identity. Social sustainability emphasizes participation, inclusion, education, healthcare, food security, and access to natural resources.

In community-based conservation, social sustainability means that all segments of the community—especially marginalized groups such as women, the elderly, or indigenous peoples—are involved in decision-making and benefit from conservation outcomes. Equitable sharing of revenues from eco-tourism, non-timber forest products, or carbon credits contributes to social cohesion and reduces conflicts.

One of the key aspects of social sustainability is the respect for indigenous and local knowledge systems. These systems are not static relics of the past, but dynamic and adaptive practices that have evolved over centuries. They often include sophisticated understanding of weather patterns, animal behavior, plant medicinal properties, and sustainable harvesting techniques. Recognizing and incorporating this knowledge strengthens conservation while affirming community identity and dignity.

Resilience—the ability to adapt to environmental, economic, or social shocks—is also a part of social sustainability. Communities that have access to diverse livelihoods, strong social networks, and institutional support are better able to cope with challenges such as climate change, resource scarcity, or disease outbreaks. Conservation programs that build such resilience help reduce dependency on destructive resource extraction and foster long-term ecological stewardship.

To achieve both ecological and social sustainability, it is important to adopt a systems approach that acknowledges feedback loops, trade-offs, and synergies. For example, promoting sustainable agriculture can enhance food security and reduce habitat loss. Similarly, restoring wetlands can improve biodiversity and protect communities from floods.

Challenges do exist, including power imbalances, lack of political will, external market pressures, and insufficient funding. But with participatory planning, transparent governance, adaptive management, and local empowerment, these challenges can be addressed.

In essence, ecological and social sustainability are not separate goals but mutually reinforcing outcomes. When communities are empowered to manage their environments in ways that reflect both ecological realities and social needs, conservation becomes a path to sustainability, equity, and resilience for people and nature alike.

Summary

1. Community-based conservation involves active participation of local communities in conserving biodiversity.
2. Local people are closely connected to ecosystems through livelihoods, culture, and traditional knowledge.
3. These communities often manage natural resources more sustainably through customary practices.
4. Community involvement leads to better protection of species, habitats, and ecological processes.
5. Empowerment, ownership, and inclusion are core principles of community-based conservation.
6. Ecological sustainability means maintaining the structure and function of ecosystems over time.
7. Social sustainability ensures communities benefit from conservation without compromising their rights or well-being.
8. Practices such as rotational grazing, regulated harvesting, and local monitoring promote sustainability.
9. Equitable benefit-sharing, respect for traditional knowledge, and inclusive governance are key.
10. Both ecological and social sustainability are interconnected and vital for long-term conservation success

Very Short Answer Type Questions

1. What is community-based conservation?
2. Define ecological sustainability.

3. What does social sustainability focus on?
4. Mention one traditional conservation practice.
5. What is benefit-sharing in conservation?
6. Name one example of ecosystem service.
7. What is habitat restoration?
8. Who are considered key stakeholders in community conservation?
9. What is a major threat to local biodiversity?
10. Name one method of sustainable resource use.

Short Answer Type Questions

1. How do local communities contribute to biodiversity conservation?
2. What is the role of traditional knowledge in conservation?
3. List two benefits of involving communities in conservation.
4. Explain the concept of ecological resilience.
5. What is meant by equitable benefit-sharing?
6. How does community participation enhance conservation outcomes?
7. Why is land tenure security important in community-based conservation?
8. Explain how cultural practices can aid biodiversity conservation.
9. What challenges can hinder social sustainability in conservation?
10. Give two examples of community-led conservation initiatives.

Long Answer Type Questions

1. Discuss the role of local communities in biodiversity conservation with examples.
2. Explain the relationship between ecological and social sustainability in community-based conservation.
3. How can traditional knowledge and practices promote ecological sustainability?
4. Describe the principles and benefits of community-based conservation.
5. What are the key components of social sustainability in conservation programs?
6. Analyze the impact of community empowerment on conservation success.
7. Discuss the importance of equitable resource access and benefit-sharing.
8. How can participatory governance support long-term conservation goals?
9. Explain the role of ecological restoration in enhancing sustainability.
10. Describe the challenges and solutions in implementing community-based conservation.

Multiple Choice Questions

1. **Community-based conservation emphasizes**

   a. Government-led initiatives only
   b. Excluding local communities
   c. Active participation of local people
   d. Fencing off protected areas

2. **Social sustainability ensures**

   a. Ignoring community welfare
   b. Focus only on wildlife
   c. Community well-being and equity
   d. No involvement of women

3. **Traditional knowledge is important because it**

   a. Is outdated and irrelevant
   b. Provides culturally rooted conservation strategies
   c. Focuses only on agriculture
   d. Promotes urban development

4. **Ecological sustainability refers to**

   a. Financial support for projects
   b. Maintaining ecosystem functions over time
   c. Growing more crops
   d. Building more infrastructure

5. **A key challenge in community conservation is**

   A. Strong cultural ties
   B. Land tenure insecurity
   C. Abundant wildlife
   D. High rainfall

6. **Benefit-sharing in conservation means**

   A. Giving all benefits to external agencies
   B. Equally distributing costs only
   C. Ensuring local communities get fair returns
   D. Removing communities from resource use

7. **Which of the following enhances ecological resilience?**

   A. Overharvesting
   B. Habitat fragmentation
   C. Rotational harvesting
   D. Pollution

8. **Community-based conservation promotes**

   A. Industrial expansion
   B. Displacement of communities
   C. Local stewardship and sustainability
   D. Ignoring cultural heritage

9. **One example of social sustainability is**

   A. Cultural erosion
   B. Access to healthcare and education
   C. Soil erosion
   D. Deforestation

10. **Sacred groves are an example of**

    A. Industrial zones
    B. Traditional biodiversity conservation
    C. Urban gardens
    D. Tourist spots

15

Agroecology and Permaculture

Agriculture has been practiced by humans for over 10,000 years. Traditional practices, such as agroecology and permaculture, have evolved into viable and efficient methods to meet dietary requirements while fostering more fertile and healthier natural and human communities. Contemporary forms of agriculture, predominantly based on the intensification of industry facilitated by fossil fuel and natural gas, are detrimental to the environment and diminishing agricultural potential through land loss, desertification, soil erosion, disruption of plant nutrient cycles, water pollution, and biodiversity loss. By utilizing renewable resources more effectively, agroecology and permaculture not only mitigate the adverse consequences of industrial agriculture but also enhance the capacity of renewable resources to sustain human and other biological systems. The implementation of agroecology and permaculture aims to provide food, energy, fiber, and habitat for the present generation without compromising the ability of natural and cultural resources to meet the needs of future generations. Consequently, all actions must be sustainable in environmental, economic, and social terms, while respecting biological and cultural diversity and human knowledge. The farm-system approach conceptualizes the farm as an ecosystem integrated within surrounding ecosystems. The management of such a system aims to minimize external inputs and maximize the cycling of mineral and biological elements within the farm. Agroecology and permaculture share common aspects but are distinct concepts. Agroecology is the discipline that studies the design and functioning of sustainable agroecosystems, whereas permaculture is the intellectual technique that applies knowledge from agroecology and other disciplines to design and establish human habitats in equilibrium with the environment. To elucidate these concepts further, their differences are discussed below.

Difference between Agroecology and Permaculture

Agroecology primarily emphasizes sustainable farming methods and is typically implemented on larger scales, such as farms or regions. It incorporates scientific research, community practices, and policy-making, focusing mainly on agricultural systems and practices.

In contrast, permaculture adopts a more comprehensive approach to creating sustainable living environments. It encompasses a broader range of practices, including agriculture, community living, and sustainable resource management. Permaculture is usually applied on smaller scales, such as individual gardens or communities, and is more practical and design-oriented, involving hands-on techniques and holistic planning.

While both agroecology and permaculture aim to promote sustainability and harmony with nature, they differ in their scope, scale, and application. Agroecology takes a wider view of agricultural systems and their ecological impacts, whereas permaculture concentrates on creating self-sustaining systems that integrate various aspects of sustainable living.

Biodiversity and Ecosystem Services

Agroecology and permaculture both significantly enhance biodiversity and ecosystem services. Agroecology promotes polyculture, agroforestry, and diverse planting to increase plant diversity, soil fertility, and pest resilience. It reduces chemical inputs, enhances soil structure, and supports water conservation through efficient infiltration and erosion control. Agroecology also fosters habitats for pollinators and natural pest predators, reducing pesticide use. Permaculture creates layered plant systems mimicking natural forests, supporting diverse species and promoting natural nutrient cycles through composting and organic matter recycling. It designs systems for efficient rainwater use and minimizes waste through recycling. Both approaches enhance carbon sequestration, soil health, water regulation, pollination, and natural pest control. Agroecology focuses on agricultural practices' ecological impacts, while permaculture integrates multiple human and natural systems for sustainable living.

Sustainable Agriculture Practices

Agroecology and permaculture advocate sustainable agricultural practices through ecological harmony, resource efficiency, and resilience. Agroecology uses diversified cropping systems, agroforestry, and polycultures to enhance biodiversity and reduce pest vulnerability. Techniques like cover cropping, reduced tillage, and organic amendments improve soil structure and productivity. Efficient water use is promoted through rainwater harvesting, contour farming, and mulching, reducing runoff and conserving water. Natural pest predators and crop rotations decrease pesticide dependence. Agroecology emphasizes resource efficiency by recycling nutrients and using local inputs. Permaculture designs self-sustaining systems integrating plants, animals, buildings, and people, maximizing resource efficiency and minimizing waste. It fosters resilient ecosystems through diverse plant growth, reducing chemical input needs, and improving soil fertility and structure with composting and organic matter. Permaculture practices encompass rainwater harvesting, swales, and greywater recycling to enhance water efficiency and conservation. Permaculture design prioritizes energy efficiency through passive solar heating, natural cooling, and renewable energy, thereby reducing external energy dependence. Nutrient loops closed via composting and recycling minimize waste, fostering sustainable and self-sufficient living. Both agroecology and permaculture advocate for harmony with natural systems, promoting resilient and sustainable agriculture. Their emphasis on ecological principles, resource efficiency, and community participation creates productive systems in harmony with the environment.

Case studies on Agroecology and Permaculture

1. **Himalaya Agroecology Research and Development Project, Nepal:** Launched in July 2022, the Himalaya Agroecology Research and Development (HARD) project in Nepal seeks to combine traditional wisdom with contemporary agroecological methods to boost food security and climate adaptability. The initiative concentrates on research, community involvement, and educational efforts. Major undertakings include setting up research areas for cash crops, creating biofertilizers, and instructing farmers in organic farming techniques. Additionally, the project empowers women through skill development and small business initiatives. The project's achievements encompass greater crop variety, enhanced soil quality, and financial gains from women-led enterprises. HARD has made substantial progress in promoting sustainable agriculture and resilient communities within the Himalayan bioregion.
2. **Bangalore Model Farm Project, India:** Situated on the outskirts of Bangalore, India, the Bangalore Model Farm is a thriving permaculture venture that emphasizes water conservation, soil health, and biodiversity. The farm maximizes yield and profitability through efficient space and resource utilization, implementing techniques such as rainwater collection and greywater reuse. Its diverse ecosystem supports a wide range of crops and livestock, enhancing resilience and sustainability. This model farm serves as an inspiring example of the economic viability and environmental advantages of permaculture practices in sustainable agriculture.

Techniques to enhance Biodiversity and Its Conservation

Currently, human habitats cover about 5% of the Earth. Anthropogenic activities contribute approximately 60% of atmospheric CO_2 and 40-45% of greenhouse gases like methane, water vapor, nitrous oxides, and halocarbons. Scientists caution that declining species diversity threatens severe consequences for the planet's biological wealth. Biodiversity is now recognized as crucial for economic and ecological stability. Initially, biodiversity loss was understood primarily as species loss, but it is now recognized that genetic loss affects much more than plant and animal populations, including crops and medicinal resources. The question of current biodiversity status has gained attention in recent decades.

The primary effect of biodiversity loss is reduced raw materials. Reduced biological diversity limits developing nations' access to indigenous plants for food and medicine, forcing reliance on foreign plant products. The moral issue of species extinction due to commercial activities is also significant. Many species, deemed insignificant, play vital ecological roles. For example, sugar-dependent ants in epiphytic plants or kelp ecosystems in temperate regions are crucial. Loss of kelp equates to losing tropical rainforests' biodiversity. Ecosystem services, essential for human society, are often provided by seemingly insignificant organisms such as insects, fungi, and forage bacteria, which are vital for nutrient extraction and decomposition.

1. Polyculture and Crop Diversification

Polyculture involves growing multiple crop species simultaneously, mimicking natural ecosystems and offering several benefits. It supports diverse wildlife, promoting ecosystem balance and resilience against pests and diseases by attracting natural predators and beneficial insects. Companion planting, a form of polyculture, includes growing mutually beneficial plants (e.g., marigolds repel nematodes and attract pollinators). Different plants improve soil structure and fertility through varying root structures and nutrient requirements, with legumes like beans and peas fixing nitrogen in the soil for other crops. Polyculture maximizes resource use by layering crops with different growth habits (e.g., tall corn shading low-growing beans).

Crop diversification includes practices like crop rotation, intercropping, and multi-cropping, preventing soil nutrient depletion, disrupting pest and disease cycles, and improving soil structure. Rotating nitrogen-fixing legumes with nitrogen-demanding cereals and growing multiple crops together enhance resource efficiency, reduce pest and disease incidence, and increase productivity. The "Three Sisters" method (corn, beans, and squash) exemplifies this. Sequential planting of fast-growing and slow-growing crops in the same field during a single season increases yield per unit area, spreads risk, and optimizes resource use.

Diverse cropping systems are more resilient to environmental stresses, such as extreme weather and climate change, reducing reliance on a single crop and enhancing food security. Diversified cropping provides multiple income sources, reducing economic risk for farmers and tapping into different market demands. Polyculture and crop diversification reduce the need for chemical inputs like pesticides and fertilizers, promoting healthier ecosystems. These practices enhance soil fertility, improve water retention, and reduce erosion, contributing to long-term sustainability. By adopting polyculture and crop diversification, farmers create more sustainable, resilient, and productive agricultural systems, benefiting both the environment and their livelihoods.

2. Agroforestry

Agroforestry is a sustainable land management practice that combines trees and shrubs with agricultural landscapes to create more productive, diverse, and ecologically balanced environments. This approach enhances soil health, improves water retention, and increases biodiversity by integrating forestry with agriculture. Agroforestry techniques include alley cropping, silvopasture, riparian buffer strips, and windbreaks, each offering unique benefits such as erosion control, habitat creation, and additional income from timber and non-timber forest products. This practice plays a crucial role in climate mitigation by sequestering carbon and reducing greenhouse gas emissions. Moreover, agroforestry promotes resilience to climate change and market fluctuations by diversifying production and providing multiple income sources. Despite requiring initial investment and knowledge, the long-term benefits of agroforestry for farmers, communities, and the environment make it a valuable approach to sustainable agriculture.

3. Habitat Restoration

Habitat restoration encompasses the process of facilitating the recovery and management of degraded, damaged, or destroyed ecosystems to their natural state. This practice aims to re-establish the ecological integrity, biodiversity, and functioning of habitats that have been adversely affected by anthropogenic activities, natural disasters, or invasive species. Habitat restoration incorporates various techniques such as reforestation, wetland reconstruction, soil rehabilitation, and eradication of invasive species. Through the reintroduction of native flora and fauna, soil stabilization, and enhancement of water quality, habitat restoration augments ecosystem services such as carbon sequestration, flood control, and pollination. The efficacy of habitat restoration often depends on a comprehensive understanding of the local ecosystem, active community participation, and long-term monitoring and maintenance. This approach not only benefits wildlife and biodiversity but also supports sustainable livelihoods and resilience against climate change, rendering it a critical component of environmental conservation and sustainable development efforts.

4. Genetic Conservation

Genetic conservation is a biodiversity conservation method that focuses on preserving the genetic diversity within and among species. It aims to maintain the genetic variation essential for the adaptability and resilience of populations in the face of environmental changes, diseases, and other threats. This method involves protecting and managing the habitats of wild species, establishing gene banks, and promoting the conservation of traditional crop varieties and livestock breeds. Gene banks store seeds, tissues, or DNA samples to ensure the availability of genetic resources for future breeding and restoration efforts. By preserving a wide range of genetic traits, genetic conservation supports the development of resilient and sustainable agricultural systems, enhances the ability of species to adapt to changing conditions, and contributes to the overall health and stability of ecosystems. Efforts in genetic conservation are crucial for safeguarding the genetic heritage of our planet and ensuring the long-term survival and well-being of both natural and cultivated species.

5. Conservation Tillage

Conservation tillage is an agricultural technique that reduces or eliminates the need for plowing, maintaining soil structure and preserving soil biodiversity. This method involves minimal soil disturbance, allowing organic matter to accumulate on the surface, which enhances soil health and fertility. By retaining crop residues on the field, conservation tillage protects the soil from erosion caused by wind and water, and improves water infiltration and retention. This technique also promotes the activity of beneficial soil organisms, such as earthworms and microbes, which contribute to nutrient cycling and soil structure. The benefits of conservation tillage extend to reducing labor and fuel costs for farmers, as well as decreasing greenhouse gas emissions by minimizing soil carbon release. Overall, conservation tillage is a sustainable farming practice that supports long-term soil health and productivity, while mitigating environmental impacts.

Very Short Answer Types Questions

1. Define Agroecology
2. What is permaculture?
3. Define polyculture
4. What do you understand by the term agroforestry?
5. Define Habitat Restoration.
6. Discuss Genetic Conservation.

Short Answer Types Questions

1. How is agroecology different from permaculture?
2. What is the impact of agroecology on biodiversity?
3. Do permaculture permits sustainable agricultural practices?
4. How agro economy contributes towards sustainable development?
5. Discuss a case study that explains the role of agroecology.
6. Discuss any two techniques that can enhance Biodiversity and conservation.
7. Discuss polyculture and crop diversification techniques and its role in promoting biodiversity.

Long Answer Type Questions

1. Discuss the role of agroecology and permaculture in promoting sustainability.
2. Discuss two differences and two similarities in the role of agroecology and permaculture.
3. Discuss case studies on agroecology and permaculture.
4. What are the major techniques that can be employed to promote biodiversity and conservation?
5. Discuss the following techniques in detail - Agroforestry, Conservation tillage and Genetic Conservation.
6. Which technique would you prefer to preserve biodiversity?

Multiple Choice Questions

1. Which technique includes practices like crop rotation and multi cropping?
   a. Polyculture
   b. Crop diversification
   c. Genetic conservation
   d. Conservation tillage
2. The agricultural technique that eliminates the need of plowing and preserving soil diversity.
   a. Crop diversification
   b. Genetic conservation
   c. Conservation tillage
   d. None of them

3. Agroforestry includes
   a. Silvopasture
   b. Riparian buffer strips
   c. None of them
   d. Both of them
4. Sequential planting of fast and slow growing crops is known as
   a.
   b.
   c.
   d. Both of them
5. Rainwater collection and grey water reuse is the technique employed in
   a. Agroecology
   b. Agroforestry
   c. Polyculture
   d. Permaculture
6. HARD stands for
   a. Himalaya endemic species research and development
   b. Himalaya agroecology Research and Development
   c. None of them
   d. Both of them
7. What are the different techniques that can help in biodiversity conservation
   a. Crop diversification
   b. Conservation tillage
   c. None of them
   d. All of them

16

Ecosystem Restoration

Introduction to Ecosystem Restoration

Ecosystem restoration is rapidly gaining recognition as one of the most essential and hopeful strategies for addressing the environmental challenges of our time. It involves the intentional process of aiding the recovery of ecosystems that have been degraded, damaged, or destroyed, whether by human activities or natural events. Unlike conservation, which focuses on protecting existing, relatively undisturbed ecosystems, restoration actively seeks to heal and rejuvenate those that have already suffered harm. This approach is fundamental to reversing biodiversity loss, mitigating climate change, and ensuring that vital ecosystem services-such as clean water, fertile soil, and climate regulation-continue to support human well-being and the health of the planet.

The scope and significance of ecosystem restoration are immense. Ecosystems are intricate networks of plants, animals, and microorganisms interacting with their physical environment. These systems can be disrupted by sudden events like deforestation, pollution, or wildfires, as well as by chronic pressures such as overgrazing, invasive species, or unsustainable farming practices. The consequences of such disturbances are profound and far-reaching. The current rate of species extinction is estimated to be dramatically higher than the natural background rate, with habitat loss and ecosystem degradation as primary drivers. This loss of biodiversity not only diminishes the richness of life on Earth but also threatens the stability and resilience of the very systems that provide essential resources for human survival.

Ecosystem restoration offers a way to halt and even reverse this degradation. Its goal is to re-establish the natural processes, functions, and interactions that sustain life, often with the vision of creating self-supporting, resilient systems capable of adapting to ongoing environmental changes. Restoration can play a significant role in preventing species extinctions and is recognized as a key strategy for addressing climate change. Restoring degraded ecosystems can contribute substantially to reducing greenhouse gas emissions, as healthy forests, wetlands, and grasslands sequester carbon and buffer the impacts of climate change.

The methods and approaches used in restoration are as varied as the ecosystems themselves. Restoration can involve removing invasive species, reintroducing native plants and animals, rehabilitating soils, restoring natural water flows, or simply removing the pressures that prevent natural recovery. Sometimes, the best approach is to let nature take the lead, allowing ecosystems to regenerate on their own once damaging activities have ceased. In other cases, active human intervention is necessary to jumpstart recovery processes that would not otherwise occur.

Restoration efforts can be initiated by governments, organizations, businesses, or local communities, and the scale can range from small, community-led projects to vast, landscape-level initiatives. Examples include reforesting degraded lands, restoring wetlands and rivers, improving agricultural practices to enhance soil health, and creating green spaces in urban areas. Each project is shaped by its unique ecological, social, and economic context, and success often depends on the active involvement and support of local communities.

The goals of restoration are evolving as our understanding of ecosystems grows. Traditionally, restoration aimed to return ecosystems to a specific historical state, often seen as pristine or untouched. However, this approach is increasingly being re-examined in light of ongoing environmental changes such as climate change and shifting species distributions. Modern restoration recognizes that it is not always possible or desirable to recreate past conditions exactly. Instead, the focus is on guiding ecosystems toward a state of health and resilience that meets both ecological and human needs in a changing world.

Restoration is a long-term process, and patience is essential. While some positive changes may be visible within a few years, the full recovery of complex ecosystems-such as mature forests or wetlands-can take decades or even centuries. The process is often iterative, involving the reintroduction of key species and the gradual return of ecological functions and interactions. Success is measured not only by the presence of native species but also by the resilience and self-sufficiency of the restored ecosystem.

It is important to emphasize that restoration is not a substitute for conservation. While restoration can achieve remarkable results, it is rarely possible to fully recreate the original complexity and diversity of lost ecosystems. Thus, restoration and conservation must go hand in hand: protecting what remains while recovering what has been lost.

The benefits of ecosystem restoration extend far beyond the environment. Healthy, restored ecosystems support a greater variety of plant and animal species, help regulate the climate, improve agricultural productivity, and protect communities from natural disasters. Restoration projects can also provide social and cultural benefits, strengthening community bonds, supporting traditional knowledge, and fostering a sense of stewardship for the land.

Economically, the advantages of restoration are substantial. Restoring degraded lands can generate significant returns in the form of ecosystem services, such as improved water quality, increased crop yields, and enhanced resilience to climate impacts. The cost of inaction-allowing degradation to continue unchecked-is much higher than the investment required for restoration

Introduction to Ecosystem Restoration

Ecosystem restoration is rapidly gaining recognition as one of the most essential and hopeful strategies for addressing the environmental challenges of our time. It involves the intentional process of aiding the recovery of ecosystems that have been degraded, damaged, or destroyed, whether by human activities or natural events. Unlike conservation, which focuses on protecting existing, relatively undisturbed ecosystems, restoration actively seeks to heal and rejuvenate those that have already suffered harm. This approach is fundamental to reversing biodiversity loss, mitigating climate change, and ensuring that vital ecosystem services-such as clean water, fertile soil, and climate regulation-continue to support human well-being and the health of the planet.

The scope and significance of ecosystem restoration are immense. Ecosystems are intricate networks of plants, animals, and microorganisms interacting with their physical environment. These systems can be disrupted by sudden events like deforestation, pollution, or wildfires, as well as by chronic pressures such as overgrazing, invasive species, or unsustainable farming practices. The consequences of such disturbances are profound and far-reaching. The current rate of species extinction is estimated to be dramatically higher than the natural background rate, with habitat loss and ecosystem degradation as primary drivers. This loss of biodiversity not only diminishes the richness of life on Earth but also threatens the stability and resilience of the very systems that provide essential resources for human survival.

Ecosystem restoration offers a way to halt and even reverse this degradation. Its goal is to re-establish the natural processes, functions, and interactions that sustain life, often with the vision of creating self-supporting, resilient systems capable of adapting to ongoing environmental changes. Restoration can play a significant role in preventing species extinctions and is recognized as a key strategy for addressing climate change. Restoring degraded ecosystems can contribute substantially to reducing greenhouse gas emissions, as healthy forests, wetlands, and grasslands sequester carbon and buffer the impacts of climate change.

The methods and approaches used in restoration are as varied as the ecosystems themselves. Restoration can involve removing invasive species, reintroducing native plants and animals, rehabilitating soils, restoring natural water flows, or simply removing the pressures that prevent natural recovery. Sometimes, the best approach is to let nature take the lead, allowing ecosystems to regenerate on their own once damaging activities have ceased. In other cases, active human intervention is necessary to jumpstart recovery processes that would not otherwise occur.

Restoration efforts can be initiated by governments, organizations, businesses, or local communities, and the scale can range from small, community-led projects to vast, landscape-level initiatives. Examples include reforesting degraded lands, restoring wetlands and rivers, improving agricultural practices to enhance soil health, and creating green spaces in urban areas. Each project is shaped by its unique ecological, social, and economic context, and success often depends on the active involvement and support of local communities.

The goals of restoration are evolving as our understanding of ecosystems grows. Traditionally, restoration aimed to return ecosystems to a specific historical state, often seen as pristine or untouched. However, this approach is increasingly being re-examined in light of ongoing environmental changes such as climate change and shifting species distributions. Modern restoration recognizes that it is not always possible or desirable to recreate past conditions exactly. Instead, the focus is on guiding ecosystems toward a state of health and resilience that meets both ecological and human needs in a changing world.

Restoration is a long-term process, and patience is essential. While some positive changes may be visible within a few years, the full recovery of complex ecosystems-such as mature forests or wetlands-can take decades or even centuries. The process is often iterative, involving the reintroduction of key species and the gradual return of ecological functions and interactions. Success is measured not only by the presence of native species but also by the resilience and self-sufficiency of the restored ecosystem.

It is important to emphasize that restoration is not a substitute for conservation. While restoration can achieve remarkable results, it is rarely possible to fully recreate the original complexity and diversity of lost ecosystems. Thus, restoration and conservation must go hand in hand: protecting what remains while recovering what has been lost.

The benefits of ecosystem restoration extend far beyond the environment. Healthy, restored ecosystems support a greater variety of plant and animal species, help regulate the climate, improve agricultural productivity, and protect communities from natural disasters. Restoration projects can also provide social and cultural benefits, strengthening community bonds, supporting traditional knowledge, and fostering a sense of stewardship for the land.

Economically, the advantages of restoration are substantial. Restoring degraded lands can generate significant returns in the form of ecosystem services, such as improved water quality, increased crop yields, and enhanced resilience to climate impacts. The cost of inaction-allowing degradation to continue unchecked-is much higher than the investment required for restoration.

Degradation of Habitats

Introduction

Habitat degradation is a significant environmental issue that threatens the diversity of life on Earth and the stability of ecosystems. Unlike complete habitat destruction, degradation refers to the gradual decline in the quality and functionality of natural environments, making them less suitable for native plants and animals. This process, often driven by human activities, disrupts the balance of nature, reduces species populations, and compromises the essential services that healthy ecosystems provide.

Causes of Habitat Degradation

Pollution

Pollution is a major cause of habitat degradation. Chemicals from agriculture, industry, and urban areas can contaminate soil and water, changing the natural balance and making habitats inhospitable for many species. For example, runoff from fertilizers can cause excessive growth of algae in water bodies, leading to oxygen depletion and harming aquatic life. Industrial pollutants and plastics can also accumulate in the environment, causing long-term harm to both land and water ecosystems.

Invasive Species

The introduction of non-native species often leads to habitat degradation. Invasive plants and animals can outcompete native species for resources, alter the structure of habitats, and disrupt natural processes. These changes can reduce biodiversity and make it difficult for native species to survive.

Over-Utilization of Natural Resources

Overgrazing, excessive logging, and unsustainable harvesting can degrade habitats by removing vegetation, compacting soil, and reducing the ability of ecosystems to recover. Such activities can lead to soil erosion, loss of plant cover, and changes in habitat structure, making it difficult for wildlife to thrive.

Alteration of Natural Processes

Human activities often disrupt natural processes such as fire, flooding, or seasonal water flow. For instance, suppressing natural fires can allow certain species to dominate, while altering river flows can impact wetlands and other water-dependent habitats.

Climate Change

Changes in climate are increasingly contributing to habitat degradation. Shifts in temperature and rainfall patterns can make habitats unsuitable for the species that live there, forcing them to move or adapt. Extreme weather events, such as droughts and floods, can further damage habitats, while rising sea levels threaten coastal environments.

Consequences of Habitat Degradation

Loss of Biodiversity

One of the most significant consequences of habitat degradation is the loss of biodiversity. As habitats become less suitable, populations of native species decline, sometimes leading to extinction. Species with specific habitat requirements are particularly vulnerable, as they have fewer options for relocation or adaptation.

Disrupted Ecological Processes

Degraded habitats often lose their ability to perform essential ecological functions, such as nutrient cycling, pollination, water purification, and soil formation. This can have widespread effects throughout the ecosystem, impacting both wildlife and human communities.

Fragmentation and Isolation

Habitat degradation often results in fragmentation, where large habitats are broken into smaller, isolated patches. These smaller areas may not be able to support healthy populations, and barriers can prevent animals from moving freely, reducing genetic diversity and increasing the risk of disease.

Reduced Ecosystem Services

Healthy habitats provide important services, including clean air and water, climate regulation, flood control, and recreational opportunities. Degradation reduces these benefits, leading to economic and social costs for people.

Examples of Habitat Degradation

1. **Grasslands and Prairies:** These areas can be degraded by the spread of unwanted plant species and by pollution from agriculture, resulting in reduced plant diversity and changes in ecosystem structure.
2. **Wetlands:** Draining and pollution can degrade wetlands, leading to declines in populations of amphibians, birds, and other wildlife.
3. **Forests:** Logging and land conversion remove critical habitat for forest-dwelling species and disrupt important ecological functions.
4. **Aquatic Ecosystems:** Rivers and streams are degraded by sedimentation, pollution, and changes in water flow, which can fragment habitats and harm aquatic life.

Strategies for Mitigating Habitat Degradation

Protected Areas and Wildlife Corridors

Creating protected areas and wildlife corridors helps preserve habitats and connect fragmented landscapes. This allows species to move, breed, and maintain healthy populations.

Ecological Restoration

Restoring degraded ecosystems through planting native vegetation, rehabilitating wetlands, and removing invasive species can help revive lost habitats and improve biodiversity.

Sustainable Land Use Practices

Using sustainable agricultural, forestry, and fishing practices can reduce pressure on natural habitats. This includes rotating crops, reducing chemical use, selective logging, and creating buffer zones to filter pollutants.

Pollution Control and Invasive Species Management

Implementing stricter pollution controls and actively managing invasive species can help restore habitat quality. Monitoring ecosystems and educating the public are important steps in preventing further degradation.

Measures to Restore Habitats

Restoring degraded habitats is essential for conserving biodiversity, supporting ecosystem services, and ensuring the long-term health of our planet. Effective restoration requires a combination of ecological, technological, and community-based approaches tailored to the specific needs of each ecosystem. Below are the principal measures and strategies used to restore habitats worldwide.

1. Reforestation and Afforestation

1. **Reforestation** involves planting trees in areas where forests have been depleted. This helps stabilize soil, improve air and water quality, and provide shelter and food for wildlife.
2. **Afforestation** is the creation of new forests in areas that were not previously forested, expanding habitat availability and enhancing carbon sequestration.

Methods

1. Planting native tree species that are resilient and attract seed dispersers.
2. Using the *framework species method*, which involves selecting 20–40 native tree species that can quickly establish a canopy, shade out invasive weeds, and facilitate natural regeneration.
3. Employing the *maximum diversity method* for smaller, isolated sites by planting a wide variety of native species to mimic natural forest composition.

2. Wetland and Aquatic Ecosystem Restoration

1. **Wetland restoration** focuses on reintroducing native plant species, restoring natural water flow, and removing barriers to improve water quality and revive aquatic habitats.
2. Rehabilitating rivers and streams by removing dams or other obstructions, allowing natural sediment flow and migration of aquatic species.

Key Actions

1. Re-establishing natural hydrology and planting wetland vegetation.
2. Monitoring water quality and controlling pollution sources.
3. Reintroducing native fish, amphibians, and aquatic plants.

3. Invasive Species Management

1. Removing invasive plant and animal species that outcompete or prey on native species is critical for habitat recovery.

2. Restoration efforts often include manual removal, controlled burns, or targeted use of herbicides and biological controls.

4. Erosion Control and Soil Stabilization

1. Preventing soil loss is vital, especially in hilly, coastal, or riverine environments.
2. Techniques include planting deep-rooted vegetation, building terraces, and using natural materials like coir logs to stabilize slopes and shorelines.

5. Reintroduction of Native Species

1. Restoring habitats often involves bringing back native plants and animals that have disappeared due to habitat loss or competition.
2. Reintroducing keystone species, such as large predators or herbivores, can trigger positive ecological cascades that benefit the entire ecosystem.

6. Creation of Protected Areas and Wildlife Corridors

1. Establishing protected areas, such as national parks and reserves, shields habitats from further degradation and allows natural processes to recover.
2. Creating wildlife corridors connects fragmented habitats, enabling animal movement, gene flow, and adaptation to environmental changes.

7. Community Engagement and Sustainable Practices

1. Involving local communities in restoration projects ensures long-term success and fosters stewardship.
2. Promoting sustainable agriculture, forestry, and fishing practices reduces pressure on natural habitats and supports restoration goals.

8. Technological Innovations in Restoration

1. **Drones and Remote Sensing:** Drones equipped with cameras and sensors can efficiently monitor large areas, assess vegetation health, and track restoration progress.
2. **GIS Mapping:** Geographic Information Systems (GIS) help identify priority areas for restoration, plan interventions, and measure outcomes.
3. **Satellite Monitoring:** Satellites provide data on habitat cover changes, guiding effective conservation actions.

9. Passive Restoration

1. Allowing nature to recover with minimal human intervention can be effective, especially where some natural regeneration potential remains.

2. Restricting harmful activities, such as grazing or logging, often leads to significant improvements in vegetation cover and habitat quality.

10. Large-Scale and Case-Specific Restoration Projects

1. **Terracing and Reforestation:** Used in degraded landscapes to slow water runoff, reduce soil erosion, and restore vegetation.
2. **Dam Removal:** Restores river ecosystems, improves fish migration, and revives riparian habitats.
3. **Coral Reef Restoration:** Involves coral seeding, selective breeding for resilience, and active transplantation to revive damaged reefs.

Principles for Effective Habitat Restoration

1. **Use of Native Species:** Always prioritize native plants and animals to ensure ecological compatibility and resilience.
2. **Adaptive Management:** Restoration projects should be flexible, incorporating ongoing monitoring and adjusting strategies as needed.
3. **Long-Term Commitment:** Ecological restoration is a gradual process requiring sustained effort, funding, and community support.
4. **Integration of Traditional Knowledge:** Local and Indigenous knowledge can enhance restoration effectiveness and cultural relevance.

Conclusion

Economic restoration is a comprehensive and essential process for rebuilding a nation's economy after a period of crisis, such as war, natural disasters, or colonial exploitation. In India's context, post-independence restoration involved addressing issues like poverty, unemployment, industrial stagnation, and agricultural decline. The establishment of planning commissions, implementation of Five-Year Plans, and adoption of mixed economy principles played vital roles in reviving the Indian economy and laying the foundation for future development.

Summary

1. **Post-Independence Challenges**: India faced poverty, food scarcity, unemployment, and industrial and agricultural backwardness.
2. **Partition Impact**: Partition led to a loss of resources, skilled labor, and markets.
3. **Mixed Economy Model**: India adopted a mixed economy with public and private sectors coexisting.
4. **Planning Commission Formation**: Set up in 1950 to formulate and implement economic plans.

5. **Five-Year Plans**: Initiated to achieve economic growth, with different plans focusing on agriculture, industry, infrastructure, etc.
6. **Land Reforms**: Abolition of zamindari and redistribution of land.
7. **Green Revolution**: Boosted agricultural productivity during the 1960s.
8. **Industrial Policy Resolutions**: Aimed to promote industrialization and reduce dependence on imports.
9. **Development of Infrastructure**: Focus on transportation, communication, and power sectors.
10. **Self-Sufficiency Goals**: Efforts to make India self-reliant in food production and industry.

Very Short Answer Type Questions

1. What does economic restoration mean?
2. When was the Planning Commission of India established?
3. Name one impact of the partition of India on the economy.
4. What kind of economy did India adopt after independence?
5. What was the main goal of the First Five-Year Plan?
6. Which revolution is associated with increased food production in India?
7. What is the role of the Planning Commission?
8. Which sector was emphasized in the Second Five-Year Plan?
9. Name one land reform undertaken after independence.
10. When did India gain independence?

Short Answer Type Questions

1. Mention any two economic problems faced by India after independence.
2. How did partition affect India's economy?
3. Explain the mixed economy system adopted by India.
4. What was the focus of the First Five-Year Plan?
5. How did the Green Revolution impact Indian agriculture?
6. What are land reforms? Give one example.
7. Describe one major objective of the Planning Commission.
8. What was the aim of the Industrial Policy Resolution of 1956?
9. How did the government try to reduce dependency on imports?
10. What role did infrastructure development play in economic restoration?

Long Answer Type Questions

1. Explain the major economic challenges India faced after independence.
2. Discuss the significance and role of the Planning Commission in India's economic development.
3. Describe the objectives and achievements of the First and Second Five-Year Plans.
4. Analyze the impact of partition on the Indian economy.
5. Discuss the main features of the mixed economy model adopted by India.
6. What were the key reforms introduced to improve agriculture post-independence?
7. Explain the role and outcome of the Green Revolution in India.
8. Evaluate the importance of land reforms in economic restoration.
9. How did industrial policies help in strengthening India's economic base?
0. Summarize India's journey towards self-sufficiency in the post-independence period.

Multiple Choice Questions

1. When was the Planning Commission of India established?

   a) 1947 **b. 1950**

   c. 1952 d. 1961

2. Which economic model was adopted by India after independence?

   a. Capitalist economy b. Socialist economy

   c. Mixed economy d. Feudal economy

3. The main aim of the First Five-Year Plan was:

   a. Industrial development **b. Agricultural development**

   c. Export promotion d. Urban planning

4. Which revolution helped increase food production in India?

   a. Blue Revolution b. White Revolution

   c. Green Revolution d. Industrial Revolution

5. What was one of the primary effects of partition on India's economy?

   a. Increase in population **b. Loss of skilled labor and markets**

   c. Industrial boom d. Agricultural surplus

6. Land reforms in post-independence India aimed to:
   a. Increase land taxes
   b. Abolish zamindari system
   c. Nationalize agriculture
   d. Introduce commercial farming
7. Which Five-Year Plan emphasized heavy industries and public sector enterprises?
   a. First
   b. Second
   c. Third
   d. Fifth
8. Which sector saw major growth due to the Green Revolution?
   a. Industry
   b. Services
   c. Agriculture
   d. Transport
9. What was the aim of the Industrial Policy Resolution of 1956?
   a. Promote privatization
   b. Encourage imports
   c. Promote industrialization and public sector
   d. Reduce exports
10. Which body was responsible for preparing India's Five-Year Plans?
   a. NITI Aayog
   b. Reserve Bank of India
   c. Planning Commission
   d. Ministry of Commerce

Section VI
Environmental Governance and Global Perspectives

17

Environmental Regulation and Compliance

Introduction

Environmental regulation and compliance have become central themes in the governance of modern societies, especially as the world grapples with the consequences of industrialization, urbanization, and population growth. The increasing awareness of environmental degradation-manifested in air and water pollution, deforestation, biodiversity loss, and climate change-has prompted nations to develop comprehensive legal frameworks to protect the environment. These frameworks are designed not only to prevent and control pollution but also to promote sustainable development, ensuring that economic growth does not come at the expense of ecological integrity. In this context, environmental laws serve as dynamic tools for balancing developmental needs with the imperative of conserving natural resources for present and future generations.

The evolution of environmental laws reflects a growing recognition that the environment is not merely a backdrop to human activity but a critical determinant of health, well-being, and economic prosperity. Environmental regulation encompasses a wide array of statutes, rules, and policies that govern how individuals, businesses, and governments interact with the natural world. Compliance with these regulations is enforced through a combination of monitoring, penalties, and incentives, with the aim of fostering responsible stewardship of the planet. This chapter explores the foundations, scope, and significance of environmental laws, highlighting their role in shaping a sustainable and equitable future.

Environmental Laws

Environmental law is a dynamic and multidimensional field that has become increasingly momentous in the modern era. It encompasses the legal frameworks, policies, and principles that govern the interaction between human activities and the natural environment. The primary objective of environmental law is to protect and preserve the environment by regulating activities that could cause harm to natural resources, ecosystems, and public health. This body of law is essential not only for preventing pollution and environmental degradation but also for promoting sustainable development-a concept that emphasizes meeting the needs of the present without compromising the ability of future generations to meet their own needs.

The Evolution and Scope of Environmental Law

Historically, environmental concerns were often overlooked in the pursuit of economic growth and industrial progress. However, the adverse concerns of unchecked development-such as air and water pollution, deforestation, loss of biodiversity, and climate change-have made it clear that environmental protection is indispensable for long-term prosperity and human well-being. The evolution of environmental law has been shaped by major environmental disasters, growing scientific understanding of ecological systems, and increasing public awareness of the interconnectedness of environmental and human health.

Environmental law covers a wide array of issues, including:

1. **Air and water quality:** Laws regulate emissions of pollutants into the air and discharges into water bodies to ensure that these vital resources remain clean and safe for use.
2. **Waste management:** Regulations govern the handling, treatment, and disposal of solid and hazardous wastes to prevent contamination and health risks.
3. **Conservation of biodiversity:** Legal provisions protect endangered species, wildlife habitats, forests, and other critical ecosystems.
4. **Land use and planning:** Environmental laws influence how land is developed, aiming to balance economic development with the preservation of natural landscapes.
5. **Climate change:** Recent legal frameworks address greenhouse gas emissions, promote renewable energy, and encourage adaptation to the impacts of global warming.

Principles Underpinning Environmental Law

Several key principles form the foundation of environmental law. These include:

1. **The Precautionary Principle:** When there is a threat of serious or irreversible environmental damage, lack of full scientific certainty should not be used as a reason to postpone preventive measures.

2. **The Polluter Pays Principle:** Those responsible for causing pollution should bear the costs of managing it to prevent damage to human health or the environment.
3. **Sustainable Development:** Development should meet the needs of the present without compromising the ability of future generations to meet their own needs.
3. **Intergenerational Equity:** The rights of future generations to a healthy environment must be considered in present-day decision-making.

These principles guide the creation and implementation of laws and policies, ensuring that environmental protection is integrated into all aspects of governance and development.

The Legal Framework in India

India has developed a robust legal framework for environmental protection, recognizing the importance of a healthy environment in its Constitution. Article 48A of the Directive Principles of State Policy directs the State to protect and improve the environment and safeguard forests and wildlife. Article 51A(g) makes it a fundamental duty of every citizen to protect and improve the natural environment.

Key Environmental Legislations in India Include

1. **The Environment (Protection) Act, 1986:** An umbrella legislation that provides the central government with broad powers to take measures for environmental protection and improvement.
2. **The Water (Prevention and Control of Pollution) Act, 1974:** Establishes mechanisms for the prevention and control of water pollution and the maintenance of water quality.
3. **The Air (Prevention and Control of Pollution) Act, 1981:** Provides for the prevention, control, and abatement of air pollution.
4. **The Forest (Conservation) Act, 1980:** Restricts the de-reservation of forests and the use of forest land for non-forest purposes.
5. **The Biological Diversity Act, 2002:** Aims to conserve biological diversity, promote sustainable use, and ensure fair sharing of benefits arising from the use of biological resources.

These laws are enforced by various regulatory agencies, including the Central Pollution Control Board (CPCB), State Pollution Control Boards, and the National Green Tribunal (NGT), which provides a specialized forum for the effective and expeditious disposal of environmental cases.

Compliance and Enforcement

Environmental compliance refers to the adherence to environmental laws, regulations, standards, and other requirements such as site permits to operate. Regulatory authorities are empowered to monitor compliance through inspections, audits, and the review of self-reported data from industries and other regulated entities. Enforcement mechanisms include the imposition of fines, suspension or revocation of licenses, closure of non-compliant facilities, and, in severe cases, criminal prosecution.

Public participation is a foundation of environmental governance. Laws often provide mechanisms for citizens and non-governmental organizations (NGOs) to participate in environmental decision-making, access information, and seek legal remedies for environmental harm. This participatory approach enhances transparency, accountability, and the effectiveness of environmental regulation.

International Dimensions of Environmental Law

Environmental issues frequently transcend national boundaries, necessitating international cooperation. Global environmental challenges such as climate change, ozone depletion, and the loss of biodiversity have prompted the development of international treaties and agreements. Notable examples include:

1. **The United Nations Framework Convention on Climate Change (UNFCCC):** Sets the framework for global efforts to address climate change, including the Kyoto Protocol and the Paris Agreement.
2. **The Convention on Biological Diversity (CBD):** Aims to conserve biological diversity, promote sustainable use, and ensure fair and equitable sharing of benefits.
3. **The Basel Convention:** Regulates the transboundary movement and disposal of hazardous wastes.

Countries are expected to incorporate their international commitments into domestic law and policy, reinforcing the global nature of environmental stewardship.

Historical Development of Environmental Legislation

Early Beginnings and Ancient Practices

The roots of environmental legislation stretch deep into human history, reflecting an enduring awareness of the need to protect natural resources and public health. Even in ancient civilizations, there were rudimentary laws and customs aimed at conserving the environment. For example, in India, rulers such as Emperor Ashoka promulgated edicts to safeguard forests and wildlife, encouraging responsible hunting and the preservation of biodiversity. Ancient Indian texts like the Vedas, Shastras, and Kautilya's Arthashastra contain references to the sustainable use of natural resources and the moral responsibility of rulers and citizens to protect the environment. These early efforts, though not codified in the modern legal sense, illustrate how societies integrated ecological conservation into daily life and governance, laying the groundwork for future legislative approaches

Similarly, in ancient Rome, as early as AD 80, the Senate passed legislation to protect the city's water supply, recognizing the importance of clean water for public health. Medieval England saw laws prohibiting the burning of coal in London and the disposal of waste into waterways, reflecting growing concerns about pollution and its impact on urban living conditions. In colonial America, leaders like William Penn mandated the preservation of forests, and Benjamin Franklin advocated for cleaner urban environments. These scattered but significant efforts reveal a long-standing, if sporadic, tradition of environmental stewardship embedded within broader legal and moral codes.

The Industrial Revolution and the Rise of Modern Environmentalism

The advent of the Industrial Revolution in the 18th and 19th centuries marked a turning point in the relationship between society and the environment. Rapid industrialization brought unprecedented economic growth but also introduced new forms of pollution and environmental degradation. In response, governments began to enact regulations to mitigate the harmful effects of industrial activities. In Britain, for instance, laws were passed to reduce air pollution from coal burning and to address the dangers posed by chemical manufacturing. However, these early regulations were often reactive and limited in scope, focusing primarily on public health rather than broader ecological concerns1.

Internationally, the pre-20th century era saw only a handful of environmental agreements, most of which were concerned with boundary waters, navigation, and fishing rights rather than pollution or ecosystem protection. Notable exceptions included conventions aimed at protecting commercially valuable species, such as the Convention for the Protection of Birds Useful to Agriculture (1902) and the Convention for the Preservation and Protection of Fur Seals (1911). These agreements, while limited, set important precedents for future international cooperation on environmental issues.

The Environmental Movement and Legislative Expansion

The mid-20th century witnessed a dramatic shift in public consciousness regarding environmental issues. The publication of Rachel Carson's *Silent Spring* in 1962 is often credited with catalyzing the modern environmental movement by highlighting the dangers of pesticides and the broader impacts of pollution on ecosystems and human health. This period saw an explosion of environmental activism, leading to the formation of green political parties, environmental NGOs, and widespread public demand for stronger environmental protections.

Governments responded by enacting a wave of comprehensive environmental legislation. In the United States, landmark laws such as the Clean Air Act, Clean Water Act, Endangered Species Act, and the establishment of the Environmental Protection Agency (EPA) fundamentally transformed the regulatory landscape. These laws expanded the scope of environmental protection from local and state levels to the national stage, introducing mechanisms for monitoring, enforcement, and public participation.

Japan, grappling with severe pollution crises in the post-war era, enacted the Basic Law for Environmental Pollution Control in 1967, one of the world's first overarching environmental statutes. Similar legislative developments occurred across Europe and other industrialized nations, reflecting a growing recognition of the need for coordinated and comprehensive approaches to environmental management.

International Environmental Law and Global Cooperation

The 1970s marked the beginning of a new era in international environmental governance. The United Nations Conference on the Human Environment, held in Stockholm in 1972, was a watershed moment, bringing together governments from around the world to address environmental challenges collectively. The resulting Stockholm Declaration articulated fundamental principles of environmental protection and established the United Nations Environment Programme (UNEP), which has since played a central role in coordinating global environmental efforts.

Following Stockholm, a series of international treaties and conventions addressed specific environmental issues. The 1973 International Convention for the Prevention of Marine Pollution from Ships (MARPOL), the 1975 Convention on International Trade in Endangered Species of Wild Fauna and Flora (CITES), and the 1987 Montreal Protocol on Substances that Deplete the Ozone Layer are just a few examples of successful international cooperation.

These agreements set standards, established monitoring mechanisms, and promoted technology transfer to address shared environmental threats.

The 1980s and 1990s saw further advancements with the Brundtland Commission's report *Our Common Future* (1987), which introduced the concept of sustainable development. The 1992 Earth Summit in Rio de Janeiro produced the Rio Declaration, Agenda 21, and three major conventions on climate change, biodiversity, and desertification, marking a new phase of integrated and holistic environmental governance.

Environmental Legislation in India: Phases of Development

India's environmental legislation has evolved through three distinct phases: ancient, pre-independence, and post-independence. The ancient phase, as discussed, was characterized by moral and customary norms rooted in religious and philosophical traditions. The pre-independence period saw the introduction of colonial laws focused primarily on resource extraction and public health, such as the Indian Forest Act (1927) and the Bengal Smoke Nuisance Act (1905).

The post-independence era marked a paradigm shift, particularly after the 1972 Stockholm Conference. India enacted a series of comprehensive laws, including the Water (Prevention and Control of Pollution) Act (1974), the Air (Prevention and Control of Pollution) Act (1981), the Environment (Protection) Act (1986), and the Biological Diversity Act (2002). These statutes established regulatory bodies, set standards for pollution control, and provided mechanisms for enforcement and public participation. The judiciary, particularly the Supreme Court of India, has played a proactive role in interpreting constitutional provisions to expand the scope of environmental protection, recognizing the right to a healthy environment as a fundamental right under Article 21 of the Constitution2.

Principles and Objectives of Environmental Laws

Environmental laws are underpinned by a set of guiding principles and objectives that shape their formulation, interpretation, and implementation. These principles serve as the philosophical and practical foundation for legal frameworks aimed at safeguarding the environment, ensuring sustainable use of natural resources, and balancing ecological integrity with developmental needs. The objectives of environmental laws are closely aligned with these principles, focusing on the prevention of environmental degradation, the promotion of sustainable development, and the protection of public health and welfare.

Fundamental Principles of Environmental Law are as Follows

1. **Sustainable Development:** Sustainable development is the cornerstone of modern environmental law. It advocates for a model of progress that fulfils the needs of the present generation without compromising the ability of future generations to meet their own needs. This principle recognizes the interdependence of environmental protection, economic growth, and social equity. Sustainable development calls for the responsible use of natural resources, reduction of waste and pollution, and the integration of environmental considerations into all aspects of policy and planning. It is enshrined in international agreements and national constitutions, such as India's guarantee of the right to a healthy environment under Article 21.
2. **The Polluter Pays Principle:** This principle asserts that those who cause environmental harm should bear the costs of managing and remedying that harm. The polluter pays principle is designed to internalize the environmental costs of economic activities, ensuring that polluters are held financially responsible for the damage they cause. It acts as a deterrent against environmental violations and encourages industries to adopt cleaner technologies and practices.
3. **The Precautionary Principle:** The precautionary principle empowers authorities to take preventive action in the face of uncertainty regarding potential environmental harm. It allows for protective measures to be implemented even when scientific evidence about risks is inconclusive. This principle is particularly relevant in managing emerging threats such as climate change, new chemicals, and genetic engineering, where waiting for full scientific certainty could result in irreversible damage.
4. **Public Trust Doctrine:** The public trust doctrine holds that certain natural resources-such as air, water, forests, and wildlife-are held in trust by the state for the benefit of the public. The state, as trustee, has a duty to protect and preserve these resources for present and future generations. This doctrine reinforces the state's responsibility to act in the public interest and prevent the over-exploitation or privatization of essential environmental assets.
5. **Intergenerational Equity:** Intergenerational equity is the principle that each generation holds the environment in trust for future generations. It emphasizes the moral and legal obligation to pass on a healthy and productive environment, ensuring that the actions of the present do not diminish the prospects of those yet to come. This principle is closely

linked to sustainable development and is reflected in both international declarations and national laws.

6. **Principle of Common but Differentiated Responsibilities (CBDR):** CBDR recognizes that while all states share the common responsibility to protect the environment, their obligations differ based on their historical contributions to environmental problems and their respective capacities. Developed countries, having contributed more to environmental degradation, are expected to take greater responsibility and provide support to developing nations. This principle is central to international environmental agreements such as the United Nations Framework Convention on Climate Change (UNFCCC).
7. **Principle of State Sovereignty and Good Neighbourliness:** States have the sovereign right to exploit their own resources, but this right is tempered by the obligation not to cause environmental harm to other states or areas beyond national jurisdiction. The principle of good neighbourliness and international cooperation encourages states to work together to address transboundary environmental issues and to resolve disputes amicably.
8. **Prevention and Rectification at Source:** The prevention principle requires that action be taken to avoid environmental damage before it occurs, rather than merely responding after the fact. The rectification at source principle mandates that environmental harm be addressed at its origin, ensuring that pollution or degradation is managed where it happens, rather than allowing it to spread or worsen5.
9. **Integration Principle:** The integration principle calls for environmental protection to be incorporated into all sectors of policy and decision-making. It recognizes that environmental considerations cannot be treated in isolation but must be mainstreamed into economic, social, and development planning to achieve sustainable outcomes.

Objectives of Environmental Laws

1. **Protection and Preservation of the Environment:** The foremost objective of environmental laws is to protect and preserve the natural environment, including air, water, soil, flora, fauna, and ecosystems. This includes preventing pollution, conserving biodiversity, and maintaining ecological balance.
2. **Sustainable Use of Natural Resources:** Environmental laws aim to ensure the judicious and equitable use of natural resources. By promoting conservation, recycling, and the use of renewable resources, these laws

seek to prevent over-exploitation and depletion of vital environmental assets689.

3. **Prevention and Control of Pollution:** A core objective is to prevent and control pollution in all its forms-air, water, soil, and noise. Environmental laws set standards for emissions, effluents, and waste management, and establish mechanisms for monitoring, enforcement, and penalties for violations.
4. **Promotion of Public Health and Welfare:** Environmental degradation has direct and indirect impacts on human health. Environmental laws are designed to protect public health by reducing exposure to hazardous substances, ensuring clean air and water, and maintaining safe living conditions.
5. **Ensuring Environmental Justice:** Environmental laws strive to ensure that all individuals and communities, regardless of their socio-economic status, have equal access to a healthy environment and are not disproportionately burdened by environmental harm. This objective is particularly important in addressing issues of environmental racism and inequity.
6. **Encouraging Public Participation and Awareness:** Effective environmental governance depends on public participation and awareness. Environmental laws often provide mechanisms for public consultation, access to information, and legal standing to challenge environmental decisions. This empowers citizens to play an active role in environmental protection.
7. **Compliance with International Obligations:** Many environmental challenges are global in nature, requiring international cooperation. Environmental laws help states fulfill their commitments under international treaties and agreements, contributing to collective efforts to address issues such as climate change, biodiversity loss, and transboundary pollution

Implementation and Enforcement of Environmental Regulations

The success of environmental laws hinges not merely on their formulation but on their effective implementation and enforcement. While the drafting of comprehensive regulations is a crucial first step, these laws must be put into practice through robust administrative systems, vigilant oversight, and a commitment to holding violators accountable. Implementation and enforcement are the vital links that transform legal principles and policy objectives into tangible improvements in environmental quality and public health.

The Process of Implementation

Implementation begins with translating legal provisions into actionable programs and procedures. This process involves the establishment of regulatory agencies, the development of technical standards, the issuance of permits and licenses, and the creation of monitoring and reporting systems. Government agencies play a central role in this phase, developing detailed rules and guidelines that clarify the obligations of individuals, businesses, and other stakeholders.

A key aspect of implementation is the dissemination of information. Regulatory bodies must educate the public, industries, and local authorities about new requirements, compliance procedures, and the importance of environmental protection. Training programs, workshops, and public awareness campaigns help ensure that all parties understand their roles and responsibilities under the law.

Resource allocation is another critical component. Successful implementation requires adequate funding, skilled personnel, laboratory facilities, and modern technology for monitoring and analysis. Governments must invest in building the capacity of regulatory agencies and providing them with the tools needed to carry out inspections, collect data, and respond to environmental incidents.

Monitoring and Compliance

Monitoring is the backbone of effective implementation. It involves regular inspections of industrial facilities, construction sites, and other regulated activities to ensure compliance with environmental standards. Monitoring may include sampling air, water, and soil, reviewing records and reports submitted by regulated entities, and using remote sensing or automated systems for real-time data collection.

Compliance is encouraged through a combination of incentives and deterrents. Incentives may include recognition programs, tax benefits, or technical assistance for companies that adopt environmentally friendly practices. Deterrents, on the other hand, involve the threat of penalties, fines, or legal action for non-compliance. The goal is to create an environment where adherence to environmental laws is the norm rather than the exception.

Enforcement Mechanisms

Enforcement is the process by which authorities ensure that environmental laws are followed and that violators are held accountable. It encompasses a range of actions, from issuing warnings and notices to imposing fines, suspending operations, or initiating criminal proceedings. The severity of enforcement measures typically depends on the nature and extent of the violation, whether it was intentional or accidental, and the potential or actual harm caused.

Administrative enforcement is the most common approach, involving actions such as issuing show-cause notices, ordering corrective measures, or suspending permits. Civil enforcement may include lawsuits to compel compliance or recover damages for environmental harm. In cases of serious or repeated violations, criminal enforcement may be pursued, leading to prosecution and, potentially, imprisonment of offenders.

Effective enforcement requires transparency, consistency, and fairness. Regulatory agencies must follow due process, allowing alleged violators an opportunity to respond to charges and present evidence in their defense. Enforcement actions should be proportionate to the offense and designed to prevent future violations.

Institutional Framework

The institutional framework for implementation and enforcement typically includes central and state environmental agencies, pollution control boards, specialized tribunals, and local authorities. These bodies are responsible for developing and updating standards, granting permits, conducting inspections, and taking enforcement actions. Coordination among these institutions is essential to avoid duplication of efforts and ensure comprehensive coverage.

In many countries, special environmental courts or tribunals have been established to handle cases involving environmental violations. These bodies provide a forum for the speedy and expert resolution of disputes, ensuring that justice is delivered efficiently and effectively.

Public Participation and Access to Justice

Public participation is a cornerstone of effective implementation and enforcement. Laws often provide mechanisms for individuals and communities to report violations, participate in decision-making, and seek legal remedies for environmental harm. Public interest litigation, citizen suits, and access to information laws empower citizens to hold both government and private actors accountable for environmental violations.

Environmental non-governmental organizations (NGOs) play a vital role in monitoring compliance, raising awareness, and advocating for stronger enforcement. Their involvement enhances transparency, accountability, and the overall effectiveness of environmental governance.

Challenges and Future Directions in Environmental Compliance

Environmental compliance is entering a period of unprecedented complexity and transformation. As societies and economies evolve, so do the challenges faced by regulators, businesses, and communities striving to meet environmental standards. The landscape is shaped by rapid industrialization, technological innovation, shifting societal expectations, and the urgent need to address global environmental crises such as climate change and biodiversity loss. Understanding the current challenges and anticipating future directions is crucial for effective environmental governance.

Major Challenges in Environmental Compliance

One of the most persistent challenges is the fragmentation and divergence of regulations across jurisdictions. As countries and regions develop their own environmental standards-often influenced by local priorities, politics, and economic conditions-businesses operating internationally must navigate a patchwork of compliance requirements. This regulatory complexity increases administrative burdens and raises the risk of non-compliance, particularly for multinational organizations that must balance global standards with local obligations5.

Financial constraints also pose significant barriers to effective compliance. Many organizations, especially small and medium-sized enterprises, struggle to allocate sufficient resources to hire environmental specialists, invest in compliance technologies, or implement robust monitoring systems. Limited budgets can lead to gaps in oversight, delayed responses to violations, and ultimately, increased environmental risk.

Technological advancement, while offering new tools for compliance, also introduces new challenges. The rapid pace of innovation in areas such as artificial intelligence, the Internet of Things (IoT), and big data analytics requires regulators and businesses to continuously update their compliance strategies. Adapting to new technologies, ensuring data privacy and cybersecurity, and integrating advanced monitoring systems can be resource-intensive and require specialized expertise.

Another pressing issue is the increasing scrutiny on environmental, social, and governance (ESG) factors. Regulatory bodies and stakeholders are demanding greater transparency, accountability, and sustainability in business operations. Companies must now demonstrate not only compliance with existing laws but also proactive efforts to minimize their environmental footprint and address social and governance issues. This shift requires the development of comprehensive ESG strategies, regular impact assessments, and transparent reporting mechanisms.

Corruption, bureaucratic inertia, and lack of coordination among agencies further complicate enforcement. In some cases, powerful economic interests resist regulation, resulting in selective or lax enforcement. Addressing these challenges requires strong leadership, sustained investment in capacity-building, and the integration of environmental considerations into broader development policies.

Opportunities and Emerging Solutions

Despite these challenges, new opportunities are emerging to strengthen environmental compliance. Technological innovations such as IoT sensors, real-time data analytics, and automated compliance systems are revolutionizing how organizations monitor emissions, track supply chain compliance, and detect violations. These tools enable more efficient, accurate, and cost-effective compliance management, reducing the risk of non-compliance and enabling proactive intervention.

Virtual audits and remote monitoring are becoming increasingly prevalent, especially in sectors like manufacturing. The integration of machine learning and predictive analytics allows for the early identification of compliance risks, while immersive technologies such as virtual and augmented reality enhance the effectiveness of remote inspections. These advances not only improve compliance outcomes but also reduce the environmental and financial costs associated with traditional, on-site audits.

The concept of "Next Generation Compliance" is gaining traction among regulators. This approach emphasizes the use of advanced monitoring technologies, electronic reporting systems, increased transparency, and innovative enforcement strategies to make compliance easier and more effective. By leveraging technology and data, regulators can streamline enforcement, reduce administrative burdens, and foster a culture of compliance.

Future Directions in Environmental Compliance

Looking ahead, several key trends are likely to shape the future of environmental compliance:

1. **Integration of ESG Criteria:** ESG considerations are becoming central to both regulatory frameworks and corporate strategies. Businesses are expected to embed sustainability, social responsibility, and ethical

governance into their operations, with regulators providing clearer guidelines and standards for ESG reporting.

2. **Adaptive Regulatory Frameworks:** As environmental challenges evolve, regulatory frameworks will need to become more flexible and adaptive. This includes updating standards to reflect new scientific knowledge, accommodating technological advances, and responding to emerging risks such as climate change, plastic pollution, and electronic waste.
3. **Global Cooperation and Harmonization:** International agreements like the Paris Agreement are setting ambitious targets for greenhouse gas reduction and sustainable development. Greater harmonization of regulations across borders, coupled with enhanced international cooperation, will be essential for addressing transboundary environmental issues and ensuring a level playing field for businesses.
4. **Environmental Justice and Inclusivity:** The principle of environmental justice is gaining prominence, emphasizing the fair treatment and meaningful involvement of all people, regardless of race, income, or nationality, in environmental governance. Future compliance efforts will need to ensure that vulnerable and marginalized communities are not disproportionately affected by environmental harm.
5. **Public Participation and Transparency:** Increased access to information, public participation in decision-making, and transparent reporting are becoming integral to effective compliance. Engaging communities and stakeholders in environmental monitoring and enforcement can enhance accountability and drive better outcomes.

Conclusion

The path forward for environmental compliance is both challenging and promising. While regulatory divergence, financial constraints, technological disruption, and enforcement hurdles present significant obstacles, the adoption of innovative technologies, the integration of ESG principles, and a commitment to transparency and inclusivity offer new avenues for progress. As environmental risks become more complex and interconnected, the future of compliance will depend on adaptive, collaborative, and technology-driven approaches that safeguard the environment for current and future generations.

Summary

1. **Environmental regulation and compliance** are essential for sustainable development and ecological balance in the face of industrialization and urbanization.
2. **Environmental laws** aim to prevent and control pollution, conserve biodiversity, and protect natural resources for present and future generations.
3. **The evolution of environmental law** has been shaped by historical events, environmental disasters, and growing scientific understanding.
4. **Key issues covered** by environmental law include air and water quality, waste management, biodiversity conservation, land use, and climate change.
5. **Foundational principles** include the precautionary principle, polluter pays principle, sustainable development, and intergenerational equity.
6. **India's legal framework** for environmental protection is robust and includes the Environment (Protection) Act, Water Act, Air Act, Forest Act, and Biological Diversity Act.
7. **Regulatory bodies** such as the Central Pollution Control Board (CPCB), State Pollution Control Boards, and the National Green Tribunal (NGT) enforce these laws.
8. **Compliance and enforcement** involve monitoring, inspections, penalties, and public participation to ensure adherence to environmental standards.
9. **International cooperation** is crucial for addressing transboundary and global environmental challenges, with treaties like the UNFCCC, CBD, and Basel Convention.
10. **Challenges** include resource constraints, technological changes, corruption, and balancing economic development with environmental protection.
11. **Future directions** focus on integrating ESG criteria, leveraging technology, ensuring environmental justice, and fostering global cooperation.

Very Short Answer Type Questions

1. What is the main goal of environmental regulation?
2. Name any two key environmental laws in India.
3. Define the precautionary principle.

4. What does the polluter pays principle mean?
5. Which body enforces environmental laws at the national level in India?
6. What is the significance of the 1972 Stockholm Conference?
7. Give one example of an international environmental treaty.
8. What is meant by environmental compliance?
9. Name one challenge in enforcing environmental laws.
10. What does ESG stand for in environmental governance?

Short Answer Type Questions

1. 65. Explain the concept of sustainable development as it relates to environmental law.
2. How does the Indian Constitution support environmental protection?
3. List and briefly describe any two principles underpinning environmental law.
4. What role does the National Green Tribunal (NGT) play in India?
5. Why is public participation important in environmental governance?
6. How has technology improved environmental compliance and enforcement?
7. What is the role of international cooperation in environmental regulation?
8. Describe the impact of the Industrial Revolution on environmental legislation.
9. How do environmental laws address the issue of biodiversity conservation?
10. What are the main functions of the Central Pollution Control Board (CPCB)?

Long Answer Type Questions

1. Discuss the historical development of environmental legislation from ancient times to the present.
2. Explain in detail the principles and objectives of environmental laws, giving relevant examples.
3. Describe the legal framework for environmental protection in India, highlighting key statutes and their enforcement mechanisms.
4. Analyze the challenges faced in the implementation and enforcement of environmental regulations.
5. Evaluate the importance of international environmental agreements and their impact on national policies.
6. Discuss the evolution of public participation in environmental governance and its significance.
7. How do environmental laws balance economic development with ecological protection? Illustrate with examples.
8. Examine the role of technology in monitoring, compliance, and enforcement of environmental laws.

9. What are the future directions for environmental compliance in the context of global environmental challenges?
10. Critically assess the effectiveness of environmental regulation and compliance in achieving sustainable development.

Multiple Choice Questions

1. What is the primary objective of environmental laws?
   a. To promote industrial growth
   b. To protect and preserve the environment
   c. To increase government revenue
   d. To support urbanization
2. Which principle states that those who cause pollution should bear the cost of managing it?
   a. Precautionary principle
   b. Polluter pays principle
   c. Sustainable development
   d. Public trust doctrine
3. Which Indian act serves as an umbrella legislation for environmental protection?
   a. Water (Prevention and Control of Pollution) Act, 1974
   b. Air (Prevention and Control of Pollution) Act, 1981
   c. Environment (Protection) Act, 1986
   d. Forest (Conservation) Act, 1980
4. Which body is responsible for enforcing environmental laws at the national level in India?
   a. Supreme Court
   b. Central Pollution Control Board (CPCB)
   c. Ministry of Finance
   d. Reserve Bank of India
5. The concept that development should meet the needs of the present without compromising future generations is known as:
   a. Intergenerational equity
   b. Precautionary principle
   c. Sustainable development
   d. Biodiversity conservation
6. Which international agreement aims to combat climate change by reducing greenhouse gas emissions?
   a. Basel Convention
   b. Paris Agreement
   c. CITES
   d. Montreal Protocol
7. What is the main function of the National Green Tribunal (NGT) in India?
   a. Oversee financial regulations
   b. Resolve environmental disputes
   c. Manage forest resources
   d. Conduct wildlife surveys

8. Which principle allows for preventive action even when scientific evidence is not conclusive?
    a. Polluter pays principle
    b. Precautionary principle
    c. Public trust doctrine
    d. Principle of state sovereignty
9. Which of the following is NOT a challenge in environmental compliance?
    a. Limited resources
    b. Technological advancements
    3. Corruption
    d. Lack of coordination
10. ESG criteria in environmental governance stand for:
    a. Environmental, Social, and Government
    b. Environmental, Social, and Governance
    c. Environmental, Safety, and Growth
    d. Environmental, Standards, and Guidelines

18

Strategies for Environmental Awareness

Introduction

Environmental awareness stands as one of the most critical pillars for achieving a sustainable future. As humanity faces unprecedented environmental challenges-including climate change, biodiversity loss, pollution, and resource depletion-the need to cultivate a deep understanding of our relationship with the natural world has never been more urgent. Environmental awareness is not merely about acquiring knowledge; it is about fostering a sense of responsibility, inspiring action, and empowering individuals and communities to make informed choices that benefit both people and the planet.

At its core, environmental awareness means recognizing the delicate balance between human activities and the environment. It involves understanding how daily choices-such as the products we buy, the energy we use, and the waste we produce-can have far-reaching consequences for natural systems. This awareness is essential for protecting vital resources like clean water, fresh air, and fertile soil, which are foundational to human health and well-being. Moreover, it helps individuals appreciate the interconnectedness of all life on Earth and the importance of preserving biodiversity for current and future generations.

The significance of environmental awareness extends beyond personal lifestyle changes. It is a catalyst for collective action and societal transformation. When communities, organizations, and governments are informed and engaged, they are better equipped to develop and implement policies and practices that safeguard the environment. This collective approach is crucial for addressing large-scale issues such as air and water pollution, deforestation, and climate change, which cannot be solved by individuals alone.

One of the most effective strategies for fostering environmental awareness is education. Schools play a pivotal role by integrating environmental topics into the curriculum, organizing hands-on projects, and encouraging outdoor learning experiences. These educational efforts help students develop critical thinking skills, a sense of environmental responsibility, and the motivation to become lifelong stewards of the Earth. Beyond the classroom, community initiatives such as clean-up drives, tree-planting campaigns, and awareness events further reinforce the importance of environmental protection and encourage active participation.

In the digital age, technology and media have become powerful tools for spreading environmental awareness. Social media campaigns, online workshops, and digital storytelling can reach vast audiences, inspiring people to adopt sustainable habits and advocate for change. Creative approaches, such as art, music, and eco-friendly challenges, engage people emotionally and make environmental messages more relatable and memorable.

Promoting environmental awareness also involves collaboration across sectors. Partnerships between schools, local governments, businesses, and environmental organizations can amplify impact and create opportunities for meaningful action. By working together, these groups can implement innovative solutions, support community-based projects, and influence policy decisions that prioritize sustainability.

Ultimately, strategies for environmental awareness must be inclusive, practical, and adaptable to different contexts. Whether through formal education, grassroots activism, or digital engagement, the goal is to build a culture where environmental stewardship is valued and practiced by all. This collective commitment is essential for ensuring a healthy, resilient planet for present and future generations.

The Origins and Objectives of Environmental Education

Origins of Environmental Education

Environmental education (EE) has a rich and evolving history, rooted in humanity's growing awareness of the need to understand and protect the natural world. The origins of EE can be traced back as early as the 18th century, when thinkers like Jean-Jacques Rousseau emphasized the importance of learning from nature. Rousseau's philosophy, articulated in his influential work Emile: or, On Education, advocated for an education that fostered a direct connection with the environment, rather than relying solely on books and abstract concepts. This early emphasis on experiential learning was echoed by naturalists such as Louis Agassiz, who encouraged his students to "study nature, not books," laying a foundation for what would later become the nature study movement.

The late 19th and early 20th centuries saw the rise of the nature study movement, particularly in the United States and Europe. This movement aimed to instill an appreciation of the natural world in children through direct observation, outdoor activities, and the use of fables and moral lessons. Anna Botsford Comstock, a leading figure in this movement, wrote the influential Handbook for Nature Study in 1911, which became a cornerstone for educators seeking to use nature as a tool for teaching both scientific concepts and cultural values.

The early 20th century also witnessed the emergence of conservation education, particularly in response to environmental crises such as the Dust Bowl and the Great Depression in the United States. Conservation education marked a shift toward a more scientific and management-oriented approach, focusing on the sustainable use of natural resources and the prevention of environmental degradation. This period laid the groundwork for a broader understanding of the interconnectedness of social, economic, and environmental systems.

The modern environmental education movement gained significant momentum in the 1960s and 1970s, fueled by growing public concern over pollution, resource depletion, and the health impacts of environmental neglect. Influential works like Rachel Carson's Silent Spring galvanized public opinion, while events such as the first Earth Day in 1970 brought environmental issues to the forefront of national and international agendas. Governments and organizations began to recognize that addressing complex environmental problems required not only scientific expertise and policy interventions but also widespread public awareness and participation.

Internationally, the importance of environmental education was solidified at the United Nations Conference on the Human Environment in Stockholm in 1972, which declared that education was a vital tool for addressing global environmental challenges. Subsequent initiatives by UNESCO and the United Nations Environment Programme (UNEP) further advanced the development and integration of environmental education into formal and informal learning systems around the world.

By the late 20th and early 21st centuries, the concept of environmental education had expanded to include themes of sustainable development, integrating social, economic, and cultural dimensions alongside ecological concerns. The United Nations Decade of Education for Sustainable Development (2005–2014) exemplified this shift, promoting a holistic approach to education that prepares individuals to contribute to a more sustainable and equitable world.

Objectives of Environmental Education

Environmental education is guided by a set of clear and interrelated objectives, all aimed at fostering a more informed, responsible, and active citizenry. These objectives reflect the evolution of EE from its early focus on nature appreciation to its current emphasis on sustainability, global citizenship, and transformative action.

1. **Awareness and Sensitivity:** The first objective of environmental education is to increase awareness and sensitivity to environmental issues. This involves helping individuals recognize the complexity and urgency of environmental problems, from local pollution to global climate change. By fostering awareness, EE encourages people to observe and reflect on their own interactions with the environment and to understand how their choices impact the world around them.
2. **Knowledge and Understanding:** EE seeks to provide learners with a sound understanding of ecological principles, environmental processes, and the interdependence of natural and human systems. This includes knowledge of topics such as biodiversity, ecosystems, resource management, and the causes and consequences of environmental degradation. A strong knowledge base enables individuals to make informed decisions and to critically evaluate environmental information presented in the media and by policymakers.
3. **Attitudes and Values:** Another central objective is to cultivate positive attitudes and values toward the environment. EE aims to instill a sense of respect, responsibility, and stewardship for the natural world. This involves fostering ethical considerations, such as fairness, justice, and the rights of future generations, and encouraging learners to adopt values that support conservation and sustainable living.
4. **Skills and Competencies:** Environmental education emphasizes the development of practical skills and competencies necessary for addressing environmental challenges. These include problem-solving, critical thinking, collaboration, and effective communication. EE often incorporates hands-on activities, fieldwork, and community projects that allow learners to apply their knowledge and skills in real-world contexts.
5. **Participation and Action:** Perhaps the most transformative objective of EE is to empower individuals and communities to participate actively in environmental protection and improvement. This means moving beyond awareness and knowledge to foster a sense of agency and responsibility. EE encourages learners to engage in actions such as conservation efforts,

advocacy, sustainable consumption, and participation in policy-making processes. The ultimate goal is to create a culture of environmental citizenship, where people are motivated and equipped to contribute to a more sustainable and just society.

6. **Lifelong Learning and Adaptability:** Environmental education is not confined to formal schooling; it is a lifelong process that adapts to changing circumstances and emerging challenges. EE promotes continuous learning, adaptability, and resilience, preparing individuals to respond to new environmental issues as they arise.

Strategies for Promoting Environmental Sustainability at Home

Promoting environmental sustainability at home is a powerful way for individuals and families to contribute to a greener, healthier planet. Small, intentional changes in daily routines and household management can collectively make a significant impact. The following strategies cover a wide range of practical actions, from reducing waste and conserving resources to making mindful purchasing decisions and embracing renewable energy.

1. Reduce, Reuse, and Recycle

One of the most fundamental strategies is to minimize waste by adopting the principles of reduce, reuse, and recycle. This involves:

1. **Reducing single-use items:** Opt for reusable bags, containers, bottles, and straws instead of disposables. Avoid products with excessive packaging and choose bulk or package-free options when possible.
2. **Reusing and repurposing:** Before discarding items, consider if they can be reused or repurposed. Glass jars, old clothing, and furniture can often find new life with a little creativity.
3. **Recycling responsibly:** Separate recyclables such as paper, glass, plastics, and metals from general waste. Familiarize yourself with local recycling guidelines to ensure materials are processed correctly.

2. Composting Organic Waste

Composting is an effective way to reduce the volume of household waste sent to landfills while creating nutrient-rich soil for gardening. Food scraps, coffee grounds, eggshells, and yard waste can all be composted. This process not only diverts organic matter from landfills, where it would produce methane, but also enriches garden soil, reducing the need for chemical fertilizers.

3. Energy Conservation

Making homes more energy-efficient is central to environmental sustainability. Key steps include:

1. **Switching to LED bulbs:** LEDs use significantly less energy and last much longer than traditional incandescent bulbs.
2. **Turning off lights and appliances:** Cultivate the habit of switching off lights and unplugging devices when not in use.
3. **Using energy-efficient appliances:** Upgrade to appliances with high energy-efficiency ratings, such as those certified by Energy Star.
4. **Improving insulation:** Seal drafts around doors and windows, add insulation to walls and attics, and use window treatments to regulate indoor temperatures.
5. **Programmable thermostats:** These devices help optimize heating and cooling schedules, reducing unnecessary energy use.

4. Water Conservation

Water is a precious resource, and conserving it at home is both environmentally and economically beneficial. Strategies include:

1. **Fixing leaks:** Even small drips from faucets or pipes can waste significant amounts of water over time.
2. **Installing low-flow fixtures:** Showerheads, faucets, and toilets designed for water efficiency can drastically reduce household water use.
3. **Shortening showers and turning off taps:** Simple habits like taking shorter showers and turning off the tap while brushing teeth or washing dishes make a difference.
4. **Collecting rainwater:** Use rain barrels to collect water for gardening and outdoor cleaning.

5. Sustainable Food Practices

Food choices have a substantial environmental impact. Sustainable practices include:

1. **Growing your own food:** Even a small vegetable or herb garden reduces reliance on produce that has been transported long distances and grown with chemicals.
2. **Reducing food waste:** Plan meals, store food properly, and use leftovers creatively to minimize waste.

3. **Composting food scraps:** As mentioned, composting is a great way to handle unavoidable food waste.
4. **Incorporating plant-based meals:** Reducing meat and dairy consumption lowers your carbon footprint and supports a more sustainable food system.

6. Mindful Purchasing and Eco-friendly Products

Being a conscious consumer is essential for sustainability:

1. **Choosing products with minimal or recyclable packaging:** Support companies that use sustainable materials and packaging.
2. **Buying second-hand or upcycled goods:** Thrift stores and online marketplaces offer furniture, clothing, and household items that reduce demand for new resources.
3. **Selecting eco-friendly cleaning products:** Homemade cleaners using vinegar, baking soda, and citrus peels are effective and avoid harmful chemicals.
4. **Opting for sustainable materials:** When renovating or redecorating, choose bamboo, cork, reclaimed wood, or recycled materials.

7. Waste Management and Zero-Waste Living

Proper waste management goes beyond recycling:

1. **Composting:** As previously discussed, this reduces landfill waste and improves soil.
2. **Repairing and mending:** Fix clothes, furniture, and appliances instead of discarding and replacing them.
3. **Donating and recycling:** Donate usable items to charity and recycle electronics and hazardous materials at designated centers.

8. Renewable Energy and Green Technology

Transitioning to renewable energy sources is a long-term investment in sustainability:

1. **Solar panels:** Installing solar panels can significantly reduce reliance on fossil fuels and lower energy bills.
2. **Solar-powered devices:** Consider solar water heaters, garden lights, and chargers.
3. **Smart home technology:** Use smart power strips and programmable thermostats to optimize energy use.

9. Creating a Healthy Indoor Environment

Indoor air quality is an often-overlooked aspect of sustainability:

1. **Houseplants:** Incorporate air-purifying plants such as snake plants or spider plants to improve indoor air quality.
2. **Non-toxic paints and finishes:** Choose paints and finishes that emit fewer volatile organic compounds (VOCs) for healthier air.
3. **Natural textiles:** Use organic cotton, bamboo, or linen for bedding and upholstery.

10. Sustainable Transportation Choices

Transportation is a significant source of household emissions:

1. **Walking and cycling:** Opt for these modes for short trips.
2. **Public transport and carpooling:** Reduce the number of vehicles on the road and lower emissions.
3. **Electric vehicles:** If possible, transition to electric or hybrid vehicles for daily commuting.

11. Involving the Whole Family and Community

Sustainability is most effective when it becomes a shared value:

1. **Education:** Teach children about sustainability through hands-on activities like gardening and recycling.
2. **Community initiatives:** Participate in local clean-up drives, tree-planting events, or community gardens.
3. **Sharing resources:** Borrow or share tools, appliances, and equipment with neighbors instead of buying new.

Advancing Environmental Education in Educational Institutions

Environmental education (EE) has become a cornerstone of preparing future generations to address the complex ecological challenges facing our world. Educational institutions-ranging from preschools to universities-play a vital role in equipping students with the knowledge, skills, and values necessary for environmental stewardship and sustainable living. Advancing EE within these settings requires a multifaceted approach that integrates curriculum innovation, experiential learning, community engagement, and the use of modern technology.

Integrating Environmental Education into the Curriculum

A fundamental strategy for advancing EE is embedding environmental concepts across subjects and grade levels. This integration ensures that environmental literacy is not treated as a standalone topic but as an essential component of science, social studies, language arts, and even mathematics. By weaving sustainability themes into diverse disciplines, students develop a holistic understanding of environmental issues and their relevance to daily life.

Curriculum development can include lessons on ecosystems, climate change, biodiversity, waste management, and renewable energy. Teachers can use real-world case studies, current events, and local environmental issues to make learning more relevant and engaging. This approach not only enhances environmental knowledge but also encourages students to think critically about the interconnectedness of natural and human systems.

Experiential and Project-Based Learning

Experiential learning is a powerful tool for deepening students' connection to the environment. Hands-on activities such as tree planting, water quality testing, composting, and biodiversity surveys allow students to apply theoretical knowledge in practical settings. Field trips to parks, nature reserves, farms, or urban green spaces provide opportunities for direct observation and foster a sense of responsibility for local ecosystems.

Project-based learning further empowers students to address real-life environmental challenges. By working collaboratively on projects-such as designing a school recycling program, restoring a habitat, or developing a sustainable garden-students build problem-solving skills, teamwork, and leadership. These experiences not only enhance academic learning but also cultivate a sense of ownership and agency in environmental stewardship.

Fostering Critical Thinking and Problem-Solving

Environmental education should go beyond rote memorization, encouraging students to analyze data, evaluate evidence, and propose solutions to complex problems. Educators can facilitate critical thinking through debates, research projects, and inquiry-based assignments. By engaging with authentic environmental dilemmas, students learn to weigh competing interests, consider ethical implications, and develop informed opinions.

This emphasis on critical thinking prepares students to navigate the uncertainties and trade-offs inherent in environmental decision-making. It also equips them with transferable skills that are valuable in a wide range of academic and professional contexts.

Building Community Partnerships

Collaboration with local communities, non-profit organizations, government agencies, and businesses can greatly enrich environmental education. Partnerships enable schools to access expertise, resources, and real-world learning opportunities beyond the classroom. For example, students might participate in citizen science projects, volunteer for local conservation initiatives, or work with municipal authorities on sustainability campaigns.

Community-based projects help students understand the relevance of environmental issues in their own neighborhoods and foster a sense of civic responsibility. They also provide opportunities for intergenerational learning, as students engage with community members and learn from their experiences and knowledge.

Incorporating Action-Oriented Learning

Effective environmental education programs include opportunities for students to take meaningful action. Service-learning, citizen science, and advocacy projects allow students to move from awareness to engagement, making tangible contributions to environmental improvement. Whether it's organizing a litter cleanup, restoring native habitats, or advocating for policy changes, action-oriented learning reinforces the idea that everyone can make a difference. These experiences often lead to sustained involvement in environmental issues, as students witness the impact of their efforts and develop a lifelong commitment to stewardship.

Leveraging Technology and Digital Tools

Modern technology offers innovative ways to enhance environmental education. Digital platforms, educational apps, and virtual reality experiences can make complex environmental concepts more accessible and engaging. For instance, students can use online tools to monitor local air or water quality, participate in global environmental campaigns, or explore virtual ecosystems.

Technology also facilitates collaboration and knowledge sharing among students, teachers, and experts across the globe. Online forums, webinars, and social media enable the exchange of ideas and best practices, building a global community of environmentally literate citizens.

Promoting Sustainable Practices Within Institutions

Educational institutions themselves can serve as models of sustainability. By adopting green practices-such as energy conservation, waste reduction, water management, and sustainable landscaping-schools demonstrate their commitment to environmental values. Involving students in these initiatives, from designing recycling systems to monitoring energy use, provides practical learning experiences and reinforces sustainable behaviors.

Eco-certification programs, such as Eco-Schools or Green Campus initiatives, offer frameworks for integrating sustainability into all aspects of school life. These programs recognize and celebrate institutions that make tangible progress toward environmental goals, inspiring others to follow suit.

Adapting to Diverse Learners and Contexts

Advancing environmental education requires sensitivity to the diverse needs and backgrounds of students. Strategies should be adapted to different age groups, learning styles, and cultural contexts. Early childhood education might focus on nature play and sensory experiences, while secondary and higher education can tackle complex scientific and policy issues. Inclusive approaches ensure that all students, including those with disabilities or from marginalized communities, have access to meaningful environmental learning.

Measuring Impact and Continuous Improvement

To ensure the effectiveness of environmental education, institutions should regularly assess learning outcomes and program impacts. Surveys, interviews, and participatory evaluation methods can provide valuable feedback from students and stakeholders. Continuous improvement, based on evidence and reflection, helps institutions refine their strategies and respond to emerging environmental challenges.

Implementing Environmental Education in the Community

Implementing environmental education in the community is a dynamic process that empowers people to understand, value, and actively protect their local environment. Unlike formal classroom settings, community-based environmental education is flexible, rooted in local realities, and often driven by grassroots participation. Its goal is to raise awareness, change behaviors, and foster a collective sense of stewardship that leads to sustainable community development.

Key Steps and Strategies

1. Community Assessment and Participation

The first step is to assess the community's specific environmental needs, challenges, and resources. This involves gathering input from local residents, leaders, and organizations to identify pressing issues-such as waste management, water quality, or habitat loss-and to understand cultural, social, and economic factors that may influence environmental attitudes and behaviors.

Active participation is crucial. Community members should be involved in every stage, from planning and leadership to implementation and evaluation. This participatory approach ensures that programs are relevant, culturally appropriate, and more likely to be embraced and sustained over time.

2. Identifying Target Groups and Tailoring Messages

Communities are diverse, with different groups-such as youth, elders, farmers, business owners, and local officials-having unique perspectives and needs. Effective environmental education tailors its messages and activities to resonate with each group. For example, workshops for farmers might focus on soil conservation, while urban residents might benefit from recycling and energy-saving campaigns.

3. Hands-On, Experiential Learning

Practical, action-oriented activities are at the heart of community environmental education. Examples include:

1. Organizing neighborhood clean-ups or tree planting events
2. Creating community gardens or composting programs
3. Conducting nature walks, water quality monitoring, or biodiversity surveys
4. Developing local recycling or waste reduction initiatives

These experiences foster a personal connection to the environment and demonstrate the tangible benefits of sustainable practices.

4. Building Partnerships

Collaboration with local organizations, schools, government agencies, and businesses expands the reach and impact of environmental education. Such partnerships provide expertise, resources, and opportunities for joint projects-like habitat restoration, citizen science initiatives, or sustainability fairs. Working together also strengthens social ties and builds a sense of shared responsibility.

5. Service Learning and Citizen Science

Involving community members in service learning and citizen science projects empowers them to contribute directly to environmental improvement.

Activities might include collecting data on local wildlife, monitoring air or water quality, or participating in conservation projects. These initiatives build scientific literacy, foster teamwork, and provide valuable information for local decision-making.

6. Utilizing Technology and Media

Digital tools and media can greatly enhance community environmental education. Social media campaigns, online workshops, and mobile apps help spread information, organize events, and engage a wider audience. Virtual tours, interactive mapping, and online citizen science platforms make learning accessible and engaging for all ages.

7. Celebrating Success and Sharing Stories

Recognizing and celebrating local achievements-such as successful clean-up drives, restored habitats, or youth leadership in sustainability-motivates continued engagement and inspires others. Sharing stories through community events, local media, and social platforms builds pride and a positive environmental identity.

Empowerment and Lasting Impact

Community-based environmental education is not just about imparting knowledge; it's about empowering people to take action and shape their environment. As community members learn together, they develop new skills, attitudes, and behaviors that support sustainability. This collective effort can lead to long-term improvements in environmental quality, health, and resilience.

Moreover, as environmental awareness grows, communities are better positioned to advocate for policies, attract resources, and influence broader change. The process of working together on environmental issues also strengthens community bonds and fosters a culture of active citizenship.

Conclusion

Environmental awareness is the cornerstone of achieving sustainability. As the world faces growing ecological challenges like climate change, pollution, and biodiversity loss, it is essential to foster environmental consciousness through education, community action, and personal responsibility. Schools, communities, and individuals must work collaboratively using technology, traditional knowledge, and policy advocacy to develop responsible behaviors. Long-term environmental protection depends on creating a culture of environmental stewardship that is informed, inclusive, and action-oriented.

Summary

1. **Environmental awareness** fosters informed decision-making and responsible behavior toward nature.
2. **Education** is a powerful tool for instilling environmental values and understanding in students.
3. The **origins of environmental education** go back to nature-study and conservation movements.
4. **Objectives** of environmental education include awareness, knowledge, values, skills, and participation.
5. **Promoting sustainability at home** includes reducing waste, saving energy, conserving water, and sustainable food habits.
6. **Renewable energy**, composting, and eco-friendly products reduce environmental impact.
7. **Educational institutions** can embed EE into the curriculum and model sustainable behavior.
8. **Experiential learning** and **project-based activities** promote hands-on environmental understanding.
9. **Community education** empowers citizens through grassroots initiatives and citizen science.
10. Collaboration between **schools, communities, and government** enhances outreach and impact.

Very Short Answer Type Questions

1. What is environmental awareness?
2. Who wrote *Silent Spring*?
3. What does EE stand for?
4. Name one benefit of composting.
5. What is meant by reduce, reuse, recycle?
6. What is the role of LED bulbs in sustainability?
7. Name one strategy to save water at home.
8. What is citizen science?
9. Give one example of a sustainable transportation method.
10. What is the full form of UNEP?

Short Answer Type Questions

1. Mention any two objectives of environmental education.
2. How can technology help spread environmental awareness?

3. What is the significance of experiential learning in EE?
4. Describe any two sustainable practices at home.
5. What is the nature study movement?
6. How does environmental awareness impact policy decisions?
7. What is the role of schools in promoting environmental sustainability?
8. Why is community involvement essential in environmental education?
9. List two ways of conserving energy at home.
10. What are eco-schools or green campuses?

Long Answer Type Questions

1. Explain the evolution and importance of environmental education.
2. Describe the objectives of environmental education with examples.
3. Discuss the strategies for promoting environmental sustainability at home.
4. How can educational institutions advance environmental education?
5. Elaborate on the role of project-based learning in fostering environmental awareness.
6. What are the key steps in implementing environmental education in the community?
7. Discuss the role of technology in environmental education and community engagement.
8. How does citizen science contribute to environmental awareness and action?
9. Describe the benefits of involving families and communities in sustainability efforts.
10. How can educational institutions model sustainability for students?

Multiple Choice Questions (MCQs)

1. Who is considered a pioneer of environmental education through the book Silent Spring?

   a. Anna Comstock b. Rachel Carson
   c. Jean-Jacques Rousseau **d. Greta Thunberg**

2. What is the main objective of environmental awareness?

   a. To increase industrial growth b. To promote consumption
   c. To inspire informed and sustainable choices d. To support political ideologies

3. Which of the following is NOT a part of 3Rs?

   a. Reduce b. Reuse
   c. Rebuild d. Recycle

4. What does composting help reduce?
   a. Energy consumption
   b. Organic waste in landfills
   c. Fossil fuel usage
   d. Rainwater runoff
5. Which educational approach involves direct observation and real-world problem solving?
   a. Rote learning
   b. Experiential learning
   c. Lecture-based learning
   d. Virtual simulation only
6. What is one benefit of planting trees in a community?
   a. Increase electricity bills
   b. Reduce biodiversity
   c. Improve air quality
   d. Decrease water retention
7. Which of the following can help reduce energy usage at home?
   a. Using halogen bulbs
   b. Keeping lights on all day
   c. Using LED lighting
   d. Running ACs with windows open
8. What is an eco-friendly alternative to chemical cleaners?
   a. Bleach
   b. Vinegar and baking soda
   c. Pesticides
   d. Ammonia
9. The Eco-Schools program encourages schools to:
   a. Build more buildings
   b. Use more paper
   c. Reduce, reuse, and recycle
   d. Increase industrial training
9. Which of the following is a sustainable transportation practice?
   a. Driving alone in a diesel SUV
   b. Walking or cycling
   c. Buying multiple cars
   d. Avoiding public transport

19

International Collaboration and Policy Advocacy

Introduction

In an increasingly interconnected world, the challenges that threaten the health of our environment, economy, and societies are no longer confined within national borders. Issues such as climate change, biodiversity loss, transboundary pollution, pandemic outbreaks, water scarcity, and trade imbalances demand collective responses that transcend geopolitical boundaries. As a result, **international collaboration and policy advocacy** have emerged as indispensable tools for nations seeking sustainable development and global harmony.

International collaboration refers to the cooperation between nations, global institutions, and regional entities in addressing shared concerns through coordinated actions, knowledge exchange, and strategic partnerships. This collaboration often involves the signing of multilateral agreements, participation in global forums, scientific cooperation, financial aid, and joint research initiatives. It is rooted in mutual benefit, where countries pool resources, technologies, and expertise to achieve goals that no single nation could accomplish alone.

Policy advocacy, on the other hand, is the active effort to influence decision-making processes, laws, and regulations at local, national, and international levels. When undertaken globally, policy advocacy involves promoting equitable, science-based, and ethically sound policies on issues such as climate justice, gender equality, sustainable development, and human rights. Civil society organizations, academic institutions, intergovernmental bodies, and even private corporations play pivotal roles in international advocacy by voicing concerns, recommending solutions, and holding governments accountable.

One of the most visible examples of international collaboration is the **United Nations Framework Convention on Climate Change (UNFCCC)**, under which countries negotiate climate action plans like the **Paris Agreement**. These frameworks illustrate how nations can collectively agree on emission targets, adaptation strategies, and financial mechanisms to mitigate global warming. Similarly, organizations such as the **World Health Organization (WHO)**, **World Trade Organization (WTO)**, **UNESCO**, and **Intergovernmental Panel on Climate Change (IPCC)** offer platforms for policy dialogue, data-sharing, and international standard-setting.

The importance of collaboration extends to **sustainable development goals (SDGs)**, a set of 17 global objectives adopted by all United Nations Member States in 2015 to end poverty, protect the planet, and ensure prosperity for all. Achieving these goals necessitates robust partnerships between developed and developing nations, especially in areas like technology transfer, capacity building, fair trade, and climate resilience.

Furthermore, global challenges often expose disparities in power, resources, and representation. **Policy advocacy helps bridge these gaps**, enabling marginalized countries and communities to have a voice in decision-making processes that affect their lives. Advocacy can involve direct negotiations, participation in international conferences, publication of position papers, and mobilization of grassroots support across countries.

Another key dimension of international collaboration is **science and technology cooperation**. Collaborative research programs in areas such as biotechnology, renewable energy, public health, and oceanography have led to major breakthroughs that benefit multiple nations. Academic institutions, through bilateral and multilateral research agreements, contribute to global knowledge generation while nurturing a culture of innovation and shared responsibility.

Moreover, global financial institutions like the **World Bank**, **International Monetary Fund (IMF)**, and regional development banks provide critical funding and policy support for countries undertaking structural reforms or sustainable infrastructure projects. However, these collaborations also raise questions about equity, conditionalities, and sovereignty, necessitating strong domestic advocacy to protect national interests.

Transboundary Environmental Challenges

Understanding Transboundary Environmental Issues

Transboundary environmental challenges refer to environmental problems that extend beyond the boundaries of a single country, affecting multiple nations and often requiring joint management and solutions. These challenges arise because natural systems-such as rivers, air currents, and migratory species-do not adhere to political borders. Consequently, pollution, resource depletion, and ecological degradation in one country can have direct and sometimes severe impacts on neighboring regions.

Major Types of Transboundary Environmental Challenges

1. **Air Pollution:** Pollutants released into the atmosphere, such as sulfur dioxide, nitrogen oxides, and particulate matter, can travel hundreds or even thousands of kilometers. Acid rain, resulting from these pollutants, can damage forests, lakes, and agricultural land far from the original source. Urban smog and industrial emissions in one country may contribute to respiratory illnesses and environmental damage in another.
2. **Water Pollution:** Many of the world's major rivers and lakes are shared by two or more countries. When upstream nations discharge untreated sewage, industrial effluents, or agricultural runoff into these water bodies, downstream countries suffer the consequences, including contaminated drinking water, loss of fisheries, and destruction of aquatic ecosystems. The management of shared water resources is often complicated by competing demands for irrigation, industry, and domestic use.
3. **Marine Pollution:** Oceans and seas are global commons, and pollution from one nation's coastline can quickly spread to others. Oil spills, plastic debris, and chemical runoff can devastate marine life, fisheries, and coastal economies across vast regions. Overfishing by one country's fleet can deplete fish stocks that are vital to the livelihoods of people in neighboring countries.
4. **Transboundary Haze and Wildfires:** In regions like Southeast Asia, large-scale land clearing and forest fires produce thick haze that drifts across borders, causing health crises, disrupting transportation, and damaging agriculture in neighboring countries. These events often require coordinated emergency responses and long-term policy interventions.
5. **Climate Change:** Perhaps the most significant transboundary challenge, climate change is driven by greenhouse gas emissions from all over the world and affects every nation. Its impacts-rising sea levels, shifting weather patterns, increased frequency of extreme events-do not respect political boundaries and necessitate a unified global response.

Socioeconomic and Health Impacts

Transboundary environmental problems can have profound effects on public health, economic stability, and social well-being. Air and water pollution are linked to respiratory diseases, cancers, and developmental disorders, while environmental degradation can undermine food security and livelihoods. The economic costs include increased healthcare expenses, loss of productivity, and damage to infrastructure and ecosystems.

Governance and Policy Complexities

Addressing transboundary challenges is inherently difficult. Countries may have different environmental standards, economic priorities, and capacities for enforcement. Disagreements over responsibility, cost-sharing, and the distribution of benefits and burdens can hinder cooperation. Effective governance requires building trust, sharing information transparently, and developing mechanisms for joint monitoring, conflict resolution, and enforcement.

Examples of Transboundary Environmental Issues

1. The Rhine River in Europe, once heavily polluted by industrial discharges, is now a model for international cooperation after decades of joint efforts to restore water quality.
2. The annual haze crisis in Southeast Asia, caused by forest fires in one country, routinely affects air quality, health, and economies across several neighboring nations.
3. The Aral Sea disaster, resulting from unsustainable water extraction by multiple countries, illustrates the catastrophic consequences of failing to manage shared resources collaboratively.

The Need for International Collaboration

Given the interconnected nature of environmental systems, no country can solve transboundary challenges alone. International collaboration is essential for sharing scientific knowledge, harmonizing regulations, and mobilizing resources for prevention, mitigation, and adaptation strategies.

International Environmental Agreements

Definition and Importance

International environmental agreements are formal treaties or conventions negotiated and adopted by two or more countries to address environmental issues that cross national boundaries. These agreements are essential tools for fostering cooperation, setting common goals, and establishing rules and standards for environmental protection at the regional or global level.

Key Elements of International Environmental Agreements

1. **Shared Objectives:** Agreements typically define clear objectives, such as reducing emissions, conserving biodiversity, or protecting shared resources.
2. **Obligations and Commitments:** Parties agree to specific actions, targets, or standards. These may be legally binding (requiring compliance) or non-binding (voluntary).
3. **Protocols and Amendments:** Many agreements are supplemented by protocols that provide more detailed rules or set new targets as scientific understanding evolves.
4. **Monitoring and Reporting:** Effective agreements include mechanisms for monitoring progress, reporting data, and verifying compliance.
5. **Dispute Resolution:** Provisions for resolving disagreements and enforcing commitments are critical for maintaining trust and effectiveness.

Examples of Major International Environmental Agreements

1. **Montreal Protocol:** Successfully phased out the production and use of substances that deplete the ozone layer, demonstrating the potential for global cooperation.
2. **Kyoto Protocol and Paris Agreement:** Established targets for reducing greenhouse gas emissions to combat climate change, with the Paris Agreement emphasizing voluntary national commitments and regular reviews.
3. **Convention on Biological Diversity:** Aims to conserve biological diversity, promote sustainable use of its components, and ensure fair sharing of benefits arising from genetic resources.
4. **Basel Convention:** Regulates the transboundary movement and disposal of hazardous wastes to protect human health and the environment.
5. **Ramsar Convention:** Focuses on the conservation and wise use of wetlands, recognizing their ecological and economic importance.

Table 19.1: Major International Environmental Agreements

Agreement/Convention	**Focus Area**	**Key outcomes/protocols**
Ramsar Convention	Wetlands protection	Conservation of wetlands of international importance
Stockholm Convention	Persistent Organic Pollutants (POPs)	Elimination/restriction of hazardous chemicals
Convention on Biological Diversity (CBD)	Biodiversity conservation	Cartagena Protocol on Biosafety
United Nations Framework Convention on Climate Change (UNFCCC)	Climate change mitigation/ adaptation	Kyoto Protocol, Paris Agreement
Convention on Long-Range Transboundary Air Pollution (CLRTAP)	Air pollution control	Gothenburg Protocol
Basel Convention	Hazardous waste management	Regulation of transboundary movements of hazardous wastes

Implementation and Challenges

Implementing international agreements is challenging due to differences in economic development, political priorities, and administrative capacities among countries. Some nations may lack the resources or technology to meet agreed targets, while others may prioritize short-term economic gains over long-term environmental protection. Ensuring compliance often requires technical assistance, financial support, and capacity-building efforts.

Role of International Organizations and Civil Society

Organizations such as the United Nations Environment Programme (UNEP), World Health Organization (WHO), and various regional bodies play vital roles in facilitating negotiations, providing scientific expertise, and supporting implementation. Non-governmental organizations (NGOs) and civil society groups also contribute by raising awareness, monitoring progress, and advocating for stronger commitments.

Successes and Limitations

Some international agreements have achieved remarkable successes. The Montreal Protocol is widely regarded as one of the most effective environmental treaties, resulting in the recovery of the ozone layer. The Convention on Long-Range Transboundary Air Pollution has led to significant reductions in harmful emissions in Europe and North America.

However, many agreements face limitations. The global response to climate change has often been criticized for lacking ambition and urgency.

Enforcement mechanisms may be weak, and political will can fluctuate with changing governments and economic conditions. In some cases, agreements are undermined by non-participation or withdrawal of key countries.

The Path Forward

To address emerging and persistent environmental challenges, international agreements must remain flexible, adaptive, and inclusive. Strengthening monitoring, enforcement, and support for developing countries is essential. Greater involvement of local communities, the private sector, and scientific institutions can enhance the effectiveness and legitimacy of these agreements.

Conclusion

Transboundary environmental challenges highlight the interconnectedness of our world and the necessity for international collaboration. International environmental agreements provide the frameworks needed to address these challenges, but their success depends on political commitment, adequate resources, and the willingness of all parties to work together for the common good. Only through sustained and coordinated global action can we ensure a healthy and sustainable environment for future generations.

Summary

1. International collaboration addresses global issues like climate change, pollution, and pandemics.
2. Policy advocacy influences global decision-making and ensures representation for all stakeholders.
3. Organizations like UNFCCC, WHO, WTO, and IPCC enable structured international cooperation.
4. The SDGs promote global partnerships for sustainable development.
5. Transboundary environmental challenges like air/water/marine pollution and climate change affect multiple countries.
6. International collaboration is essential for managing shared ecosystems and resources.
7. International environmental agreements set shared goals and obligations.
8. Treaties like the Montreal Protocol, Paris Agreement, and Basel Convention guide global environmental action.
9. Implementation of agreements faces challenges like economic disparities and lack of political will.

10. Civil society and international organizations play a key role in awareness, monitoring, and support.

Very Short Answer Type Questions

1. What is international collaboration?
2. Name one transboundary environmental issue.
3. What does SDG stand for?
4. Which treaty deals with ozone layer protection?
5. What is the full form of WHO?
6. Which agreement succeeded the Kyoto Protocol?
7. Define policy advocacy.
8. What is the focus of the Basel Convention?
9. Name one agency that supports environmental agreements.
10. What does the Paris Agreement aim to reduce?

Short Answer Type Questions

1. Define policy advocacy with an example.
2. List any two goals of international collaboration.
3. How do transboundary environmental issues arise?
4. What are the main features of international environmental agreements?
5. Why climate change is considered a transboundary issue?
6. Mention any two examples of successful international environmental agreements.
7. What role do NGOs play in global policy advocacy?
8. Briefly describe the role of the UNFCCC.
9. How does water pollution affect international relations?
10. What is the Cartagena Protocol?

Long Answer Type Questions

1. Explain the importance of international collaboration in addressing global environmental problems.
2. Discuss the main features and outcomes of the Paris Agreement.
3. Describe various types of transboundary environmental challenges with examples.
4. What are the components of a successful international environmental agreement?
5. How do international organizations facilitate global cooperation?
6. What challenges are faced during the implementation of international agreements?
7. Compare the Montreal Protocol and the Kyoto Protocol.
8. Evaluate the role of civil society in international policy advocacy.

9. Explain the significance of the SDGs in promoting global partnerships.
10. How can countries overcome governance challenges in transboundary issues?

Multiple Choice Questions

1. Which international treaty deals with ozone-depleting substances?
   a. Kyoto Protocol
   b. Montreal Protocol
   c. Paris Agreement
   d. Cartagena Protocol
2. What does the Paris Agreement aim to limit global warming to?
   a. 3°C
   b. 2.5°C
   c. 2°C
   d. 1.5°C
3. Which organization plays a key role in global health issues?
   a. WTO
   b. WHO
   c. IMF
   d. UNEP
4. The Ramsar Convention is related to:
   a. Air pollution
   b. Hazardous waste
   c. Wetland protection
   d. Climate change
5. The SDGs were adopted in which year?
   a. 2012
   b. 2015
   c. 2018
   d. 2020
6. The Kyoto Protocol mainly addressed:
   a. Wetland conservation
   b. Transboundary water issues
   c. Greenhouse gas emissions
   d. Airborne diseases
7. The term "transboundary pollution" refers to pollution that:
   a. Occurs within a single country
   b. Spreads to other countries
   c. Happens in oceans only
   d. Affects only industries
8. Which agreement includes the Cartagena Protocol?
   a. Ramsar Convention
   b. Basel Convention
   c. Convention on Biological Diversity
   d. Stockholm Convention
9. The UNEP is responsible for:
   a. Health policy
   b. Environmental leadership
   c. Human rights protection
   d. Currency monitoring
10. What is a major obstacle to the enforcement of international agreements?
    a. Climate patterns
    b. Financial profits
    c. Political differences and economic inequality
    d. Cultural traditions

References

Archer D. How it went down last time. Nature Geoscience. 2010 Dec;3(12):819-20.

Bairey E, Kelsic ED, Kishony R. High-order species interactions shape ecosystem diversity. Nature communications. 2016 Aug 2;7(1):12285.

Bathiany S, Bastiaansen R, Bastos A, Blaschke L, Lever J, Loriani S, De Keersmaecker W, Dorigo W, Milenković M, Senf C, Smith T. Ecosystem resilience monitoring and early warning using Earth observation data: challenges and outlook. Surveys in Geophysics. 2024 May 3:1-37.

Begon M, Townsend CR. Ecology: from individuals to ecosystems. John Wiley & Sons; 2020 Nov 17.

Berner RA. The carbon cycle and carbon dioxide over Phanerozoic time: the role of land plants. Philosophical Transactions of the Royal Society of London. Series B: Biological Sciences. 1998 Jan 29;353(1365):75-82.

Berryman AA. Principles of population dynamics and their application. Garland Science; 2020 Nov 25.

Bodenheimer FS. Problems of animal ecology. (No Title). 1938.

Butler JH, Montzka SA. The NOAA annual greenhouse gas index (AGGI). NOAA Earth System Research Laboratory. 2016;58:55-75.

Campbell MO. Biogeochemistry and the Environment. Springer; 2024.

Canfield DE, Glazer AN, Falkowski PG. The evolution and future of Earth's nitrogen cycle. science. 2010 Oct 8;330(6001):192-6.

Chapin III FS, Matson PA, Vitousek P. Principles of terrestrial ecosystem ecology. Springer Science & Business Media; 2011 Sep 2.

Chaudhary S, Sindhu SS, Dhanker R, Kumari A. Microbes-mediated sulphur cycling in soil: Impact on soil fertility, crop production and environmental sustainability. Microbiological Research. 2023 Jun 1;271:127340.

Cohen JE. Food webs and niche space Princeton University Press. Princeton, 237p. 1978.

Cooper N, Brady E, Steen H, Bryce R. Aesthetic and spiritual values of ecosystems: Recognising the ontological and axiological plurality of cultural ecosystem 'services'. Ecosystem Services. 2016 Oct 1;21:218-29.

Courchamp F, Dunne JA, Le Maho Y, May RM, Thebaud C, Hochberg ME. Fundamental ecology is fundamental. Trends in ecology & evolution. 2015 Jan 1;30(1):9-16.

Dantzer B, Mabry KE, Bernhardt JR, Cox RM, Francis CD, Ghalambor CK, Hoke KL, Jha S, Ketterson E, Levis NA, McCain KM. Understanding organisms using ecological observatory networks. Integrative Organismal Biology. 2023;5(1):obad036.

Delmas E, Besson M, Brice MH, Burkle LA, Dalla Riva GV, Fortin MJ, Gravel D, Guimarães Jr PR, Hembry DH, Newman EA, Olesen JM. Analysing ecological networks of species interactions. Biological Reviews. 2019 Feb;94(1):16-36.

Dodd AN, Kusakina J, Hall A, Gould PD, Hanaoka M. The circadian regulation of photosynthesis. Photosynthesis research. 2014 Feb;119:181-90.

Ehrlich HL, Oremland RS, Zehr JP. Biogeochemical cycles. e LS. 2001 May 30.

Elton C. Animal ecology University of Chicago Press. Chicago, Illinois. 1927.

Fahey JW, Zalcmann AT, Talalay P. The chemical diversity and distribution of glucosinolates and isothiocyanates among plants. Phytochemistry. 2001 Jan 1;56(1):5-1.

Fath BD, Killian MC. The relevance of ecological pyramids in community assemblages. Ecological Modelling. 2007 Nov 10;208(2-4):286-94.

Friederichs K. A definition of ecology and some thoughts about basic concepts. Ecology. 1958 Jan 1;39(1):154-9.

Fuentes-Lara LO, Medrano-Macías J, Pérez-Labrada F, Rivas-Martínez EN, García-Enciso EL, González-Morales S, Juárez-Maldonado A, Rincón-Sánchez F, Benavides-Mendoza A. From elemental sulfur to hydrogen sulfide in agricultural soils and plants. Molecules. 2019 Jun 19;24(12):2282.

Gann GD, McDonald T, Walder B, Aronson J, Nelson CR, Jonson J, Hallett JG, Eisenberg C, Guariguata MR, Liu J, Hua F. International principles and standards for the practice of ecological restoration. Restoration Ecology. 27 (S1): S1-S46.. 2019;27(S1):S1-46.

Grimm AM. Interannual climate variability in South America: impacts on seasonal precipitation, extreme events, and possible effects of climate change. Stochastic Environmental Research and Risk Assessment. 2011 May;25:537-54.

Günal S, Hardman R, Kopriva S, Mueller JW. Sulfation pathways from red to green. Journal of Biological Chemistry. 2019 Aug 16;294(33):12293-312.

Hairston Jr NG, Hairston Sr NG. Cause-effect relationships in energy flow, trophic structure, and interspecific interactions. The American Naturalist. 1993 Sep 1;142(3):379-411.

Hamel C, Landry C, Elmi A, Liu A, Spedding T. Nutrient dynamics: Utilizing biotic-abiotic interactions for improved management of agricultural soils. Journal of Crop Improvement. 2004 Sep 21;11(1-2):209-48.

Hernes T, Weik E. Organization as process: Drawing a line between endogenous and exogenous views. Scandinavian journal of management. 2007 Sep 1;23(3):251-64.

Huntley BJ. General Concepts in Ecology. InEcology of Angola: Terrestrial Biomes and Ecoregions 2023 Mar 8 (pp. 185-214). Cham: Springer International Publishing.

Ings TC, Montoya JM, Bascompte J, Blüthgen N, Brown L, Dormann CF, Edwards F, Figueroa D, Jacob U, Jones JI, Lauridsen RB. Ecological networks–beyond food webs. Journal of animal ecology. 2009 Jan;78(1):253-69.

Jakhar S. Fundamentals of Ecology. Techsar Pvt Ltd; 2024 Apr 20.

Jax K. Ecological units: definitions and application. The quarterly review of biology. 2006 Sep;81(3):237-58.

Jayaramaiah RH, Egidi E, Macdonald CA, Singh BK. Linking biodiversity and biotic interactions to ecosystem functioning. Journal of Sustainable Agriculture and Environment. 2024 Sep;3(3):e12119.

Kader AA, Saltveit ME. Respiration and gas exchange. Postharvest physiology and pathology of vegetables. 2002 Dec 4:31-56.

Keller DR, Golley FB, editors. The philosophy of ecology: from science to synthesis. University of Georgia Press; 2000.

Kobziar LN, Hiers JK, Belcher CM, Bond WJ, Enquist CA, Loudermilk EL, Miesel JR, O'Brien JJ, Pausas JG, Hood S, Keane R. Principles of fire ecology. Fire Ecology. 2024 Apr 25;20(1):39.

Koning EA. The three institutionalisms and institutional dynamics: Understanding endogenous and exogenous change. Journal of Public Policy. 2016 Dec;36(4):639-64.

Küppers EU. Operational Principles of Nature: Universal Development Tools of Long-Term Proven Biodiversity-Rich Management. InIngenious Principles of Nature: Do We Reckon With Nature Or Nature Reckons With Us 2022 Oct 27 (pp. 219-299). Wiesbaden: Springer Fachmedien Wiesbaden.

Lang JM, Benbow ME. Species Interactions and.

Lawrence JR, Germida JJ. Enumeration of sulfur-oxidizing populations in Saskatchewan agricultural soils. Canadian Journal of Soil Science. 1991 Feb 1;71(1):127-36.

Lecours V, Devillers R, Schneider DC, Lucieer VL, Brown CJ, Edinger EN. Spatial scale and geographic context in benthic habitat mapping: review and future directions. Marine Ecology Progress Series. 2015 Sep 15;535:259-84.

Lindahl BD, Tunlid A. Ectomycorrhizal fungi–potential organic matter decomposers, yet not saprotrophs. New Phytologist. 2015 Mar;205(4):1443-7.

Lindeman RL. The trophic-dynamic aspect of ecology. Bulletin of mathematical biology. 1991 Mar;53:167-91.

Liu YX, Dou SJ, Li M, Wang XL. Production of indole and hydrogen sulfide by the oxygen-tolerant mutant strain Clostridium sp. Aeroto-AUH-JLC108 contributes to form a hypoxic microenvironment. Archives of Microbiology. 2022 Aug;204(8):486.

Liu YX, Dou SJ, Li M, Wang XL. Production of indole and hydrogen sulfide by the oxygen-tolerant mutant strain Clostridium sp. Aeroto-AUH-JLC108 contributes to form a hypoxic microenvironment. Archives of Microbiology. 2022 Aug;204(8):486.

Maier RM. Biogeochemical cycling. InEnvironmental microbiology 2015 Jan 1 (pp. 339-373). Academic Press.

May R. Stability and Complexity in Model Ecosystems. Princeton, NJ: Prineeto.

McCauley DJ, Gellner G, Martinez ND, Williams RJ, Sandin SA, Micheli F, Mumby PJ, McCann KS. On the prevalence and dynamics of inverted trophic pyramids and otherwise top-heavy communities. Ecology Letters. 2018 Mar;21(3):439-54.

Miguez NG, Mason BM, Qiu J, Cao H, Callaghan CT. Urban greenspaces benefit both human utility and biodiversity. Urban Forestry & Urban Greening. 2025 May 1;107:128791.

Molles, Manuel C., and Barry W. Barker. "Ecology: concepts and applications." (1999).

Moore JC, de Ruiter PC, McCann KS, editors. Food Webs. Cambridge University Press; 2018.

Morin PJ. Community ecology. John Wiley & Sons; 2009 Apr 13.

Nelson RH. Ecological science as a creation story. The Independent Review. 2010 Apr 1;14(4):513-34.

Niknahad-Gharmakher H, Piutti S, Machet JM, Benizri E, Recous S. Mineralization-immobilization of sulphur in a soil during decomposition of plant residues of varied chemical composition and S content. Plant and soil. 2012 Nov;360:391-404.

O'Brien EM. Biological relativity to water–energy dynamics. Journal of biogeography. 2006 Nov;33(11):1868-88.

Odum EP, Barrett GW. Fundamentals of ecology.

Ontiveros VJ, Capitán JA, Casamayor EO, Alonso D. The characteristic time of ecological communities. Ecology. 2021 Feb;102(2):e03247.

Oren A. Biogeochemical cycles. eLS. 2001:1-0.

Peccerillo A. Geochemical Cycles—The Circulatory System of Planet Earth. InAir, Water, Earth, Fire: How the System Earth Works 2021 Oct 4 (pp. 129-152). Cham: Springer International Publishing.

Piccolo JJ, Taylor B, Washington H, Kopnina H, Gray J, Alberro H, Orlikowska E. "Nature's contributions to people" and peoples' moral obligations to nature. Biological Conservation. 2022 Jun 1;270:109572.

Pirgmaier E. The value of value theory for ecological economics. Ecological Economics. 2021 Jan 1;179:106790.

Popolizio T. A Photosynthesis Primer. Understanding Organisms: An Evolutionary, Ecological and Comparative Approach. 2024 Oct 7.

Pringle RM, Hutchinson MC. Resolving food-web structure. Annual Review of Ecology, Evolution, and Systematics. 2020 Nov 2;51(1):55-80.

Rahn H. The role of N_2 gas in various biological processes with particular reference to the lung. Studies in Pulmonary Physiology: Mechanics, Chemistry, and Circulation of the Lung. 1963;1:208.

Rana SV. Essentials of ecology and environmental science. PHI Learning Pvt. Ltd.; 2013 May 13.

Rascio N, LA ROCCA N. Biological nitrogen fixation. InReference module in earth systems and environmental sciences 2013 (pp. 1-16). Esevier.

Rockwood LL. Introduction to population ecology. John Wiley & Sons; 2015 Apr 7.

—saed. Nature is beyond value because we are part of nature. Capitalism Nature Socialism. 2019 Apr 3;30(2):143-56.

Saha B, Saha S, Roy PD, Padhan D, Pati S, Hazra GC. Microbial transformation of sulphur: an approach to combat the sulphur deficiencies in agricultural soils. Role of rhizospheric microbes in soil: Volume 2: nutrient management and crop improvement. 2018:77-97.

Sepie AJ. More than stories, more than myths: Animal/human/nature (s) in traditional ecological worldviews. Humanities. 2017 Oct 20;6(4):78.

Sfara E, El-Hani CN. Ecosystem health and malfunctions: an organisational perspective. Biology & Philosophy. 2023 Oct;38(5):37.

Shennan C. Biotic interactions, ecological knowledge and agriculture. Philosophical Transactions of the Royal Society B: Biological Sciences. 2008 Feb 27;363(1492):717-39.

Singh V. The Ecosystems. InTextbook of Environment and Ecology 2024 Mar 23 (pp. 17-40). Singapore: Springer Nature Singapore.

Slobodkin LB. Ecological energy relationships at the population level. The American Naturalist. 1960 May 1;94(876):213-36.

Smith RL, Smith TM. Elements of ecology. Pearson Higher Ed; 2015 Jul 27.

Stein LY, Klotz MG. The nitrogen cycle. Current Biology. 2016 Feb 8;26(3):R94-8.

Strickland TC, Fitzgerald JW, Ash JT, Swank WT. Organic sulfur transformations and sulfur pool sizes in soil and litter from a southern Appalachian hardwood forest. Soil science. 1987 Jun 1;143(6):453-8.

Thompson RM, Brose U, Dunne JA, Hall RO, Hladyz S, Kitching RL, Martinez ND, Rantala H, Romanuk TN, Stouffer DB, Tylianakis JM. Food webs: reconciling the structure and function of biodiversity. Trends in ecology & evolution. 2012 Dec 1;27(12):689-97.

Trebilco R, Baum JK, Salomon AK, Dulvy NK. Ecosystem ecology: size-based constraints on the pyramids of life. Trends in ecology & evolution. 2013 Jul 1;28(7):423-31.

Turner MG, Gardner RH, Turner MG, Gardner RH. Organisms and landscape pattern. Landscape ecology in theory and practice: pattern and process. 2015:229-85.

Turney S, Buddle CM. Pyramids of species richness: the determinants and distribution of species diversity across trophic levels. Oikos. 2016 Sep;125(9):1224-32.

Ussiri DA, Lal R. Carbon sequestration for climate change mitigation and adaptation. Cham, Switzerland: Springer International Publishing; 2017 Mar 28.

Vandegehuchte ML, de la Pena E, Bonte D. Relative importance of biotic and abiotic soil components to plant growth and insect herbivore population dynamics. PLoS One. 2010 Sep 23;5(9):e12937.

Vázquez DP, Gianoli E, Morris WF, Bozinovic F. Ecological and evolutionary impacts of changing climatic variability. Biological Reviews. 2017 Feb;92(1):22-42.

Vellend M. Conceptual synthesis in community ecology. The Quarterly review of biology. 2010 Jun;85(2):183-206.

Wang D, Xu PY, An BW, Guo QP. Urban green infrastructure: Bridging biodiversity conservation and sustainable urban development through adaptive management approach. Frontiers in Ecology and Evolution. 2024 Aug 1;12:1440477.

Wang H, Lv G, Cai Y, Zhang X, Jiang L, Yang X. Determining the effects of biotic and abiotic factors on the ecosystem multifunctionality in a desert-oasis ecotone. Ecological Indicators. 2021 Sep 1;128:107830.

Wang N, Zhao K, Li F, Peng H, Lu Y, Zhang L, Pan J, Jiang S, Chen A, Yan B, Luo L. Characteristics of carbon, nitrogen, phosphorus and sulfur cycling genes, microbial community metabolism and key influencing factors during composting process supplemented with biochar and biogas residue. Bioresource Technology. 2022 Dec 1;366:128224.

Watanabe M, Takahashi A, Kojima H, Miyata N, Fukui M. Desulfofustis limnaeus sp. nov., a freshwater sulfate-reducing bacterium isolated from marsh soil. Archives of Microbiology. 2022 Oct;204(10):647.

Widdel F, Bak F. Gram-negative mesophilic sulfate-reducing bacteria. InThe prokaryotes: a handbook on the biology of bacteria: ecophysiology, isolation, identification, applications 1992 (pp. 3352-3378). New York, NY: Springer New York.

Williams DW, Fuester RW, Metterhouse WW, Balaam RJ, Bullock RH, Chianese RJ, Reardon RC. Density, size, and mortality of egg masses in New Jersey populations of the gypsy moth (Lepidoptera: Lymantriidae). Environmental Entomology. 1990 Aug 1;19(4):943-8.

Wyngaard N, Cabrera ML. Measuring and estimating sulfur mineralization potential in soils amended with poultry litter or inorganic fertilizer. Biology and Fertility of Soils. 2015 Jul;51:545-52.

Xing S, Leahy L, Ashton LA, Kitching RL, Bonebrake TC, Scheffers BR. Ecological patterns and processes in the vertical dimension of terrestrial ecosystems. Journal of Animal Ecology. 2023 Mar;92(3):538-51.

Xu HB, Tsukuda M, Takahara Y, Sato T, Gu JD, Katayama Y. Lithoautotrophical oxidation of elemental sulfur by fungi including Fusarium solani isolated from sandstone Angkor temples. International Biodeterioration & Biodegradation. 2018 Jan 1;126:95-102.

Yasin G, Nawaz MF, Rasool F, Yousuf MT, Nazir MQ, Javed A. Effect of Canopy Cover of Capparis decidua Forsk. on Soil Conditions in Thal Desert. Journal of Bioresource Management. 2016;3(2):3.

Zhao C, Wang J, Zang F, Tang W, Dong G, Nan Z. Both water content and sulfur-oxidizing bacterial community affect elemental sulfur oxidation in soils with different textures. Available at SSRN 4029206. 2022.

Index